The Wild Life of Our Bodies

ROB DUNN

THE WILD LIFE
OF OUR BODIES

PREDATORS, PARASITES, AND PARTNERS THAT SHAPE WHO WE ARE TODAY

HARPER

An Imprint of HarperCollins*Publishers*
www.harpercollins.com

HarperCollins books may be purchased for educational, business, or sales promotional use. For information, please write: Special Markets Department, HarperCollins Publishers, 10 East 53rd Street, New York, NY 10022.

FIRST EDITION

Designed by Fritz Metsch

Library of Congress Cataloging-in-Publication Data

Dunn, Rob R.
  The wild life of our bodies : predators, parasites, and partners that shape who we are today / Rob Dunn. — 1st ed.
      p. cm.
  Includes bibliographical references.
  ISBN 978-0-06-180648-3
  1. Microbial ecology. 2. Human ecology. 3. Human evolution. 4. Host-parasite relationships. I. Title.
  QR171.D86 2011
  579'.17—dc22   2010043564

11 12 13 14 15   OV/RRD   10 9 8 7 6

for Monica, my favorite wild life

Contents

Introduction

Some night, when the moon sneaks through your curtains and finds you still awake in bed, look beside you at your companion. (If you are alone, look at yourself.) Look at the fingernails, smooth beside the rougher skin, not unlike claws. Look at the hands, full of bones strung together by strings of tendons. Follow the bones just below the skin of the arm to the elbow and up along the beautiful shoulder to the neck that, at this moment, might seem to be the loveliest thing you have ever encountered. This body, composed of flesh and desires, evolved in the trees in Africa and Asia, where those nails helped cling to a branch to keep from falling to predators on the ground. You find yourself at this moment beside an animal that was, very recently, wild.

Some days we remember and feel our connection to what came before us. As we watch a chimpanzee on TV and see its gestures, kindnesses, and cruelties, we feel empathy. As we pick up a turtle in the road, we notice its legs, strange eyes, and a body not so unlike ours. We feel it moving in our hands like some deep muscle of life. But most days we are less aware of being part of a broader community of living species. We no longer see ourselves as part of nature.

Yet our history clings to us, whether we notice or not. In the last several years, dozens of new and separate discoveries by researchers in anthropology, medicine, neurobiology, architecture, and ecology—especially ecology—have made that much clear. The more we distance ourselves from our evolutionary history, the more we seem to feel the pull of uncut strings of our heritage. There are metaphorical or even spiritual ways in which we might ache for the past, but I mean something much more physical. I mean the ache that our bodies feel in being removed from the ecological context in which they existed for millennia. In being separated from the web of life with which we evolved, we are feeling effects, some good, others bad, but nearly all consequential, not just for how but even who we are.

We take our modern rituals of work and leisure for granted, and yet for nearly every bit of our history we lived outside, naked or nearly so. We sat, when we dared, on branches. We slept in nests made of sticks, mud, and smooth leaves. We roamed and foraged and knew about the landscape that was around us because we had to, because from its colorful fruits and treasures we ate and either lived or did not. In our transition to modern life, one can make long lists of the things our bodies might miss. It was not long ago that we still walked on four legs. Our bodies remain awkward standing straight. We run fast, but not so fast, and we do it by leaning forward toward that older gait. Our backs, as we sit all day, every day, pain us with our four-footed history. And as the eminent scientist Paul Ehrlich, author of *The Population Bomb*, put it, standing up also made it harder for us to sniff each other. So much for the good old days.

Biologists and philosophers have pondered for generations the ways in which our modern lives may be disconnected from our pasts, out of synch. We are haunted by this dissonance, as many have acknowledged, but what seems to have been relatively missed is the origin of the ghosts. They arise from changes in the species with which we interact. When you look beside you in bed, you notice no more than one animal (alternative lifestyles and cats notwithstanding). For nearly all of our history, our beds and lives were shared by multitudes. Live in a mud-walled hut in the Amazon, and bats will sleep above you, spiders beside you, the dog and cat not far away, and then there are the insects beating themselves stupid against the dwindling animal-fat flame. Somewhere near you, perhaps hanging in the palm roof, would be the drying herbs of medicine, a cooked and salted monkey hanging on a stick, and whatever else is necessary, all gathered or killed, all local, all touched and held and known by name. In addition, your gut would be filled with intestinal worms, your body covered in multitudes of unnamed microbes, and your lungs occupied by a fungus uniquely your own. Beyond the edge of the village, past the darkness between houses, would be an even wilder nature filled with insects flexing and scraping their parts together in song, trees falling to the ground, bats fighting among the fruit, and then, of course, the predators who walk silently along our paths, waiting to pounce.

So it is that the biggest difference between our modern human lives and the way we used to live is not the difference of housing styles and convenience (the transition from outhouse to penthouse). It is instead the change in our web of ecological connections. We have gone from lives immersed in nature to lives in which nature appears to have disappeared. Our disconnection from the nature in which we evolved is unprecedented in its extent and in its consequences.

We may love our new way of living, the bright lights and clean counters, the delicious food and the air-conditioning—at least our conscious brains may. Meanwhile, our bodies continue to act as though they expect to meet our old companions, the species with which they tangled, generation upon generation, for tens of millions of years. Some of the ways we have distanced ourselves from other species are good; I do not miss smallpox. Other changes are neutral. They affect who we are, but not necessarily for the better or the worse. Many changes, though, are clearly bad. In recent years, for example, a new suite of diseases has begun to plague us. Sickle cell anemia; diabetes; autism; allergies; many anxiety disorders; autoimmune diseases; preeclampsia; tooth, jaw, and vision problems; and even heart disease are all becoming more common. More and more, these modern problems seem to be the consequence of changes not in levels of pollution, globalization, or even health care systems, but instead of changes in the species we interact with. It is not that we have lost particular species as much as that we have tried to remove whole kinds of life—parasites, bacteria, wild nuts and fruits, and predators, to name a few. The loss of intestinal worms seems to be making many of our bodies ill, just as the circuits in our brains that evolved to deal with predators are now causing us to lose our minds. Our conscious brains have led us to clean our lives of the rest of nature, but the rest of our body, from our guts to our immune system, is having second thoughts.

The researchers studying different aspects of our disconnection from nature are in different fields. They do not tend to talk to one another, yet they have come to parallel conclusions about the extent of the consequences of our disconnection. An immunologist holds up our intestines and sees the consequences of having removed our worms. An evolutionary biologist looks at the appendix and notices what it had been doing in our bodies all along without being noticed. A primatologist looks at the neurons in our brain and sees the vestiges of predators. Psychologists look at our fears of strangers and our wars and see in them a mark, a kind of stigma, of changes in our exposure to disease. Each thinks they have discovered something important. They have—here I attempt to bring these stories together, weaving through them the common reality that our past haunts us. As I do, I try to step back to reveal the elephant in the room, or rather the effects of having removed the elephant from the room, along with worms, microbes, birds, fruit, and the rest of the most readily apparent life.

We all know about the biodiversity crisis, but the related crises resulting from changes in the kind of nature we interact with is similarly immediate. Whether lying in bed or sitting in front of your computer, when you ache, you ache with the history of your origin. You ache with the context you miss. The savannas and forests of our ancestry are still with you. They come to you, like the pain of a missing limb, when you sneeze, when your back aches, or when you are scared. They even come to you each time you choose what to plant, eat, or buy. This history comes to some more than others, but in one way or another, it comes to us all.

In the pages that follow, I tell the story of the consequences of our changing relationships with the rest of nature. I begin with our parasites and then discuss, in turn, the species we depend on directly (our mutualists), our predators, and then our diseases. I conclude by considering the crossroads at which we find ourselves. We have options. One, the one we are headed toward, is a world in which our daily lives are more removed from nature (which is itself increasingly impoverished) and we are sicker, less happy, and more anxiety-ridden for it. In this world, we treat our problems with more and more medicines in an attempt to use chemicals to restore what we miss from other species. We live in a bubble from which we look out at the rest of life. The other options are more radical, but no less possible. Through the stories of a handful of half-wild visionaries, I will consider some of these radical options that include giant living buildings, predators in our cities, and the restoration of parasitic worms to our guts' wild plains.

In the end, what we need in our daily lives is not quite wilderness. Wilderness is what we did away with to allow ourselves to live free of malaria, dengue, cholera, and large carnivores eating our loved ones. We need a nature managed so as to complement our happy lives, a kind of wildness, perhaps. It is taboo to say that we should manage the nature closest to us for us, but ever since we first started to farm or control pests that is what we have always done. The step we must take now is to manage with more care and nuance. We can favor good bacteria in our mouths, and discourage bad bacteria. We have just chosen not to. We can introduce harmless nematodes into our bodies to restore our immune system. We can expose ourselves to the species in which we find joy, curiosity, and happiness. We can even, more ambitiously, create green cities,

cities more revolutionary than just buildings with green rooftops, cities in which entire walls are built out of life. Imagine butterflies emerging from cocoons on flowers growing out of high-rise apartment balconies. Imagine predators diving on prey on street corners—hawks in Manhattan, bears in Fairbanks. Imagine all the species—or if not all of them, more of them—and their wild calls, back outside our doors.

In the last century, we used antibiotics to kill all of the bacteria in our guts in order to get rid of a single problematic species. It was the century in which we killed all of the insects in our fields in order to control the few pest species. It was the century in which we killed wolves everywhere to save sheep in some places. It was the century in which we scrubbed our counters clean to "get rid of germs." All these actions saved tremendous numbers of lives but also left us with new more chronic problems and a nature devoid of its richness. We know more now and can act more wisely to create for ourselves more natural and healthier lives. The solution to the problems caused by our "clean living" is not as simple as just playing in the dirt. Our task is to create a new kind of living world around ourselves, one that we interact with in many different ways, a living world that is not just the species that survive deforestation, antibiotics, and disturbance, but instead some more intelligent and lush garden.

Let our lives again be where the wild things are.

PART I

Who We All Used to Be

I

The Origins of Humans
and the Control of Nature

In the summer of 1992, Tim White saw the remains that changed his life. The first thing he saw was a tooth, a single molar. And then as he approached the spot in the clay bed, there was more. He could not be sure what he was looking at. They could have been the remains of a dog almost as easily as those of a teenage girl. He could not even be sure whether there was just one body or several. A search party was staged and every bit of potential evidence began to be collected. Soon, a little farther away, other clues were discovered—more teeth, and an arm bone. The flesh was long gone, yet in their precise geography, these parts seemed to tell a story.

White stepped back from the bones and walked around them to gain perspective. The more he looked, the more he was able to sort out what he was seeing. But it took time. It was not until 1994, two years later, that enough bones turned up to reconstruct the body, or at least more of its parts. Ultimately, several individuals would be discovered, but it was this first one that called to him. All these years removed from her last breath, she still commanded attention. He could scarcely look away. She stirred a feeling in him—maybe it was the heat mixing with his ego, a kind of psychological indigestion—yet he began to imagine it was something else. Every scientist who studies fossils hopes that one day his walk in the desert will be interrupted by a find everyone else missed, a find so important that the desert itself seems to increase in worth. With time, White began to believe that this was what had happened to him.[1]

Tim White, a professor of biological anthropology at the University of California, Berkeley, has been working with the bones of human ancestors and other primates for decades. He knows the bones of monkeys, apes, and men as intimately as anyone knows anything. He has run his fingers over millions of bones, drawn them, tapped them, dug them out. Time and intuition suggested to White that these bones in the sand were not quite a woman. Nor were they quite an ape. White could not prove where they belonged on the tree of life, not as they lay disordered in the desert, but he felt in some deep and primitive part of his brain that they were significant. Not the missing link connecting humans and apes, but something more. Perhaps they were the bones that made the entire search for a missing link irrelevant. So much of fossil work has to do with native intuition, sorting the ordinary from the extraordinary upon a quick glance or a feel. White's gut knew this was extraordinary. The skull was unusual. The feet were unusual. And when White and his colleagues looked at the sediment in which they were found, it was a thin layer sandwiched between two volcanic events, events of known ages, between which played out the life of their quarry, a life whose date of birth was 4.4 million years ago.[2] The bones had been left there long before the origin of humans or that famous fossil Lucy, on which so much of our existing understanding hinged. If White was right, this find would immortalize him. If he was wrong, well, he might be just one more anthropologist left half mad in the dust of his own imagination.

Certainly there were things that pointed to White's madness. The odds of finding a fossil as unique and important as he thought this one might be were extraordinarily low, a billion to one, if not worse. Yet, if White was looking for affirmation, he could also find it. The context of this discovery alone suggested he could be on to something. He and his colleagues were working in Ethiopia's Afar desert. Their site, called Aramis, was not far from a place where other early-hominid bones had been found in 1974. Nor was it far from where he and colleagues had discovered the very earliest bones of humans, some 160,000 years ancient.[3] If White was going to excavate these bones, he wanted to do it right. "Right," though, is expensive in both time and money. The temptation to do it quickly, to make a surgical but dirty strike, would have been great. He resisted. Credibility in the study of human evolutionary history is hard to come by but easy to lose. What would come next—the many tiny bones and fragments of bones, each one picked from the ground, treated, and pieced together slowly and carefully—would have to be done perfectly. A single fragment of jaw would come to occupy months of an anthropologist's time. A shard of pelvis, weeks more. And there were just so many bones. It seemed as if this body had been trampled on by ancient hippos, only to be punished a little more each year by the grinding movement of the earth, the tunneling of termites and ants and, more simply and less forgivingly, the passage of time.[*] These bones had 4.4 million years to fall apart. He hoped it would not take quite that long to put them back together. All of Tim White's assistants and all of his colleagues struggled. It was not just that the bones had been smashed to pieces. The pieces themselves were brittle. When handled incautiously, they would turn to dust. A few did.

One hopes for a breakthrough, a great and leaping moment of "Aha!" None came. White published a small paper on the find in 1994, more to spray his territory than as a revelation.[4] At that point, nothing yet seemed done. What seemed particularly unresolved was the broader story of who these bones belonged to— what she ate, how she moved, and, more generally, how she lived. White and his colleagues would have to have all the bones in place to see that. Once they did, they would be able to compare this skeleton to other younger ones and, of course, to their own bodies. What White and company wanted to see were the differ-

---

*White thinks it was actually trampled by hippos, literally.

ences. Some things in particular would be telling: the size of the skull and hence the brain, the shape of the hips and thus how this woman walked, and the feet. (It could be said that biological anthropologists have a thing for feet; the point of a toe can mean the difference between a foot that clings to a branch and one that sprints.) Nor were the intricate bones all that White and his crew sought. They also gathered the other fossils they found around this woman, all of them—other animals, even the remains of plants. They wanted to see this whole world for what it was, whatever that might be. Jamie Shreeve, a *National Geographic* editor, has described White as being "hard and thin as a jackal,"[5] but maybe he is more like a hyena, an animal that gathers all that it can from each broken-down piece of bone.

White and his team scarcely talked to anyone about what they were doing. No one outside the group knew exactly what had been discovered. Details were leaked one year to the next, but the details seemed to conflict, almost as though false clues were being left intentionally. Meanwhile, what White was beginning to think was that the woman in the sand—Ardi, as he would affectionately come to call her—was the earliest complete skeleton of a human ancestor.[6] If so, hers would arguably be the most important hominid fossil ever discovered. This was enough to keep White ardently at his work. In fact, ardent does not begin to be a strong enough word.

As White and his team worked, it was clear that the bones they were assembling looked, in many ways, human. The differences between what White and his team had found and the bones of modern humans were, in the broader context of evolution, tiny. She may have been 4.4 million years old, but much of her was like a human child. The same would have been true for her organs and cells, had they lasted. She was like us for the simple reason that the main features of our bodies evolved far earlier than the earliest hominid or even the earliest primate. To find the bones of animals with much different parts, you must go far deeper into the layers of dirt. By the time Ardi was born, we were almost completely who we are today, minus a few bells and whistles, or perhaps better said, big brains, tools, and words.

Most of our parts evolved in some context not only different from that in which we use them today but different even from that in which the fossil woman discovered by White would have used them. We share nearly all our genes with chimpanzees and, even more, Tim White would come to argue, with the bearer of the bones he discovered. But we also share most of our traits and genes with fruit flies, a fact upon which modern genetics depends for its succor and funding. We even have many genes in common with most bacteria, genes that exist in each of our cells.

The layer in which Tim White was studying his fossil find was, at its deepest, about two feet beneath the surface of the desert sand and sediment. Two feet is the depth of sediment that built up across 4.4 million years, sometimes a few grains at a time, sometimes more. The layers of sediment in which fossils and history are trapped are not laid down evenly, but if they were, the layer in which the story of life begins would be nearly half a mile in the earth. At the bottom of that sand pile, one can find the era of the first living cell. Already it was a little bit like each of us. It had genes that we still have, genes necessary for the basic parts of any cell. Between that moment and Ardi was the origin of the mitochondria, the tiny organs in our cells that render energy from non-energy, the first nucleus in a cell, the first multicellular organisms, and the first backbone. When primates show up, just thirty feet below the surface, the depth of a well, they were small, runty even, and, no offense, not very smart, but they were already nearly identical to us genetically.

When the individual that White found had evolved, our hearts had been beating, our immune systems had been fighting, our joints clicking and clacking, and our parts otherwise being tested in our vertebrate ancestors against the environment for several hundred million years. Across these vast stretches of time, climates waxed and waned, continents moved against each other. Yet a few realities remained unperturbed by these machinations of dirt and sky. The sun rose and fell. Gravity pulled every action and inaction to the earth. Parasites attached themselves. No animal has ever been free of them. Predators ate everything; no animal has ever been free of them either. The pathogens that cause disease were common, though perhaps less predictably present than parasites and predators. Every species existed in mutual dependency with other species, in relationships that evolved essentially with the origin of life. No species was an island. No species had ever, in all of that time, gone it alone.

All these things were true not just across most of Ardi's life, or most of primate evolution, but since the very first microbial cells evolved and another cell realized the possibility of taking advantage of them. The interactions among species are life's gravity, predictable and weighty. Beginning in the layers of earth in which Tim White was digging, or perhaps slightly more recently, these interactions would begin to change. For the first time in the entire history of life, our lineage began to distance itself from other species on which it had once depended. This change would make us human. We were not the first species to use tools or to have big brains. We were not even the first species to be able to use language. But once we had big brains, language, culture, and tools, we were the first species that set out to systematically (and at least partially consciously) change the biological world. We favored some species over others and did so each place we raised a home or planted a field. Anthropologists have been arguing for a hundred years about what makes a modern human, but the answer is unambiguous. We are human because we chose to try to take control. We became human when the earth and all of its living things began to look like wet clay, when our hands, meaty with flesh, began to look like tools.

When five years had passed and Tim White still had not published any more results from his find, rumors circulated that he had gone a little mad. One can imagine the scenario. After piecing together thousands of bones, White could have easily become obsessed with going back to find those last missing pieces out in the sand. So White might have dug and dug until he spent his life out in the desert, in a hole. Then, in 2009, Tim White came out of his hole and submitted, along with his tribe of colleagues, eleven separate papers to the prestigious scientific journal *Science*, all of which were published. In the papers, White and his colleagues introduced the young female *Ardipithecus ramidus* they called Ardi. To White, it was as if he had made Ardi and her kin. She stood at about four feet. Her nose was flat, and in the reconstruction, she gazes permanently ahead. Her fingers are long and her big toe sticks out to the side like a thumb. She was not quite beautiful and yet to White she was lovely.

When the results were published, Ardi was on the front pages of newspapers around the world, always looking out wide-eyed, as if she had just been surprised. White may or may not have been immortalized, but Ardi was. *National Geographic* prepared a full-color series on her. She is the new Lucy, though both older and, at least in White's telling, more significant. Her body seemed to be an ancestor of our lineage or at the very least close kin, and she is unlike anything else that has ever been found. She seems to have traits, splayed toes for example, for walking four-legged among trees, and other traits for walking two-legged on the ground, although even that much is speculative. What is not speculative is that these bones are the most complete reconstruction of an early humanlike creature.

Nor are her circumstances debatable. She was found among other bones and evidence that, when pieced together, clearly show that she and her kin were living in a damp, tropical woodland, not a desert. Based on the animal bones and other evidence found around her, there would have been antelopes, monkeys, and palm trees. Ardi's bones indicate that they were nourished on figs and other fruits and nuts, but also some meat, both of insects and other animals. She would have once stood on a branch not far from where White found her, nibbling at figs and perhaps even wondering about her place in the broader scheme of things.* She used sticks as tools to help her eat when she was hungry, but she had no fire, no stone tools. She had not yet tried to take control of the land. She was like the other species, still wild, still covered in microbes and worms, and still more likely to die in a large cat's mouth than of old age.

With White's publications, Ardi went from unknown to famous in a remarkably short time. It is unknown where Ardi's reassembled remains will end up. In the standard arrangement, she would be placed in the lineup of our ancestors, the one that starts out with a microbe or a fish and then culminates with a man typing on a computer. In such an arrangement, Ardi would be presented looking forward. Given, though, that she was found with her bones pointed in many directions, it isn't any more right or wrong to think of her as lying on top of her own (and our) long history and looking up from that point of view. She would stare up at the shallow sand above her. In those few feet of dirty history modern humans evolved. As they did, the enduring presence of parasites, pathogens, predators, and mutualists was about to change, for the very first time.

Initially, the layers of sediment and bone laid down over Ardi's body were essentially unchanged from the one in which she was born and died. The forests persisted for generations, replete with monkeys and palms. It took 2 million years for big changes to happen. By the time the grains of those years had fallen over Ardi, the first tools were being made by our ancestors, perhaps her descendants. They were crude—pounding rocks, sharp-edged stones,

---

*Although the fruit on the tree of knowledge is now often cast as an apple, early biologists had great fun discussing other possibilities. The great namer of life Carl Linnaeus suggested the banana, since its shape is just a tiny bit sexual, as was Linnaeus's easily titillated mind. Others suggested figs, like those that Ardi ate, a more appealing possibility to my mind. Buried in each successful fig is a dead wasp, the pollinator who remains, a symbol of the interactions among species on which every sweetness depends, the interactions we have changed.

scrapers, and diggers—but useful and used. Ardi was a million years deep before the next stage began. It was a stage during which hominids such as *Homo erectus,* who used these crude tools, would give way to those who used hand axes—larger blades with a teardrop shape—to chop up bodies, though perhaps still not yet to kill them. Amazingly, six more inches of sand would accumulate, 500,000 years, before anything really changed. Across these generations, hand axes were made a 100,000 times in as many places, nearly always in exactly the same way.

Two hundred thousand years ago, with just an inch or so of sand left to accumulate before the modern age, Neanderthals and early humans began to tie their stones to sticks. Tying stones to sticks was brilliant, at least from the perspective of our ability to kill other animals. When you had to run up to a lion and hit it with a hand ax, the odds were stacked against success. But with a stick attached to that sharp rock, the odds looked at least a little bit better. One imagines that there was, when our ancestors first figured out how to tie sticks to rocks, a high demand for long sticks. These tools were clumsy but served their purpose. With them, we began to kill animals, many of them. Their bones piled up in our early caves, but we had not yet caused the extinction of any other species. We were still just one species among many, although starting to get some attitude and starting to see, perhaps, the possibility of getting some more.

Twenty-eight thousand years ago, all that was left to lay down was a layer of sand and sediment as thin as powdered sugar. In that sprinkling of time would come everything else that has happened to us—you, me, and the rest of humans. If we want to look for what makes us different as humans, it comes in this slice of time during which Neanderthals, that last holdout of what were once many species of hominids, went extinct. Twenty-eight thousand years ago, we found religion. Stone beads start to accumulate in the sediment, as do grave sites. Statues of women with big butts and breasts become all the rage, early evidence that old preferences have a way of repeating themselves, or perhaps they never go away. We developed a more "sophisticated" culture, and as we did, we began to take control of the land. The moment that made us human in that series of happenings was not the language, the gods, or even the ability to draw Rubenesque women in stone. It was when we decided that when a leopard stalked the cave, we ought to go after it and kill it. When we decided to kill a species not for food or in self-defense, but instead in order to control what lived and did not live around us, when we did that, we were then fully human.

The extent to which we have changed the earth around us in the meager years we have been a species is astounding, but may have been inevitable, a consequence of our attempts, however bumbling, to survive. The ability to kill animals with pointy stones and sticks changed us, as did fire. We burned to cook. We burned millions of acres, but crudely. We burned forests and grasslands without particular preference. We burned what would ignite, when we pleased. The abilities to build our own dwellings, kill large animals, and transform landscapes with fire combined with a peripatetic urge that would come to transform not just parts of tropical Africa and Asia, but the world. Humans arrived in Australia roughly 50,000 years ago, and not long after, all the biggest animals went extinct. Humans arrived in the New World 20,000 to 13,000 years ago, and with their arrival, mastodons, mammoths, dire wolves, saber-toothed tigers, and more than seventy other large mammal species went extinct.

The extinctions of the megafauna of Australia and the Americas were hardly the end of the story. As our populations grew denser, we outstripped the ability of the land to provide for us through meat, nuts, and fruit alone. What had long been a kind of informal planting of favorite things became more formal. We tamed plants and then also wild beasts, cows, pigs, goats, and more. Farming arose and spread. With farming, our lifestyles transformed and our impacts magnified. We burned lands to make them clear for farming. We killed the wild animals that might compete with our cows and goats.

In addition to all of our many intentional effects, we also wrought unintended effects. Among them were those brought about by the species we carried with us from place to place. Some of those species—pigs and goats and chickens—were things we brought, like fire, to make each new place more like the last one. Others species were accidental, stowaways that were either invisible or sneaky. Rats moved with us, as did flies. Species that could not live with us went extinct. Spared were only those species resistant to our spears and fire, and then some of those species went extinct owing to the rats, pigs, goats, or one of the other species we were moving around.

After each of these changes, we made the world different from what it had been. We did so by making simple changes that favored whole habitats and suites of species that our eyes perceived as good, and disfavoring species we thought were bad. In essence, we created a few new kinds of habitats that we then re-created wherever we went. All of this continued at increasing rates, as populations expanded and our ability to invent new tools increased. Bigger guns let us kill more things faster. DDT let us kill pests from planes. Antibiotics let us kill bacteria. This killing became more necessary as we changed our landscapes. Without it, diseases had become rampant in our new, more populous centers. Without it, pests had grown thick on our monocultures of food. Without the killing, everything we have achieved would revert to the entangled bank we started in, and so we kneel and spray.

Forty years ago, when people wrote about Lucy, they described her lifestyle as primitive. Now, further into our experiment as modern humans, when one looks at Ardi and her lifestyle (or for that matter Lucy) it is hard not to use the word "idyllic," perhaps as a function of a change in perspective about our own "success." Four million years ago, life in Aramis, Ethiopia, was, of course, not idyllic. Yet elements of Ardi's simple, besieged life can seem, if not good, whole, as though they fit together, each piece of her ecological puzzle connected with its counterpiece. Ardi lived as animals have always lived, with parasites, predators, and little control of the rest of nature. She picked at fleas and dreamed of the leopard's footsteps. We live today in vast areas rebuilt by our own hands to exclude predators, to grow our few grasses (wheat, corn, rye) instead of forests; areas where pests, parasites, and pathogens are cleaned away. We have lived like this for just a tiny slice of history, a scratch of footprints in the loosest sand. In living this way, one can see us from two perspectives. From a great distance, we still look tiny before the magnitude of nature. Up close though, one sees something of the opposite. We have exerted incredible control over nature. We have warmed the entire earth, even as it rotates and circles the sun. We have tried to take control in order to improve our lot, but that control has brought us to a relationship with the rest of the living world far different from that which any species has ever lived.

Right now, you are at almost no risk of predation. No tigers lurk in your kitchen or yard. You are at a low risk of encountering a parasite. But you are also likely to struggle to see, around you in your life, anything resembling a wilderness devoid of the impact of humans. These realities have consequences, more than we have realized. You might call them side effects, except that they seem to be right in front of us, knocking on our door. They are the ghosts of our ecological history. They knock softly but carry the weight of life's billions of years.

PART II

Why We Sometimes Need Worms and Whether or Not You Should Rewild Your Gut

2

When Good Bodies
Go Bad (and Why)

We expect few things more eagerly than progress, progress since Ardi, but also since yesterday. Among the simplest measures of our progress is the quality and length of our lives. It was not long ago that we were covered in hair and could expect to live fewer than forty short years between birth and predation. By the turn of this last century, life expectancies in developed countries had inched past eighty years. For most (though not all, a point to which we will return) of human history we have lived longer than the generation that preceded us. In 1850, life expectancy in the United States was forty years, in 1900 forty-eight years, in 1930 sixty years, and so on, and it was easy to imagine it would go on like this forever, with each generation living longer than the one preceding it.[1] That is, it was easy to imagine until recently, when projections of life expectancies in many of the "civilized" parts of the world began to plateau or even, in some places, decline in longevity and also, some might argue, quality.[2] In the wealthiest countries our older, healthier, happier future is coming into question. Our children may expect to live more afflicted and perhaps even shorter lives than we do. That much is clear. Less clear is why. Here then is a murder mystery in which we are nearly all the victims.

We should be living longer, healthier lives. We have figured out ways to kill more and more of the species that once tried to live at our expense. If some creature clambers into your orifices or through your skin, there is a pill for it, a spray, or maybe a salve. Got germs? Use an antibacterial wipe. Got tapeworms? Take a pill. Most of our long-standing ills can be remedied, at least with enough money. But just as we seem to be getting better at ridding ourselves of the old threats, a set of "new" diseases—including Crohn's, inflammatory bowel disease, rheumatoid arthritis, lupus, diabetes, multiple sclerosis, schizophrenia, and autism, among others—has become more and more common, and these diseases appear to be, at least in part, what is plaguing us. These diseases, contrary to our standard ideas about progress, have become most common precisely in those countries where we spend the most on health care and public health. Whether American, Belgian, Japanese, or Chilean, we in the "modern world" are getting sick in new ways.

One can imagine many reasons for why people in developed countries might suffer from problems that those in developing countries do not. Nearly everything that differs between developed and developing countries might be the culprit. The difference could be pollution, pesticides, or "in the water." It could be diet or our social interactions. Beginning between 1900 and 1950 and continuing on until the present day, a variety of these new diseases, many of them autoimmune and allergic in nature, have become more and more common. During this same time, nearly everything about our lives changed. We began to travel more. We vacuumed instead of swept. We started living in suburbs. Toothpaste with fluoride came into common usage, as eventually did pogo sticks, nose-hair clippers, double lattes, electronic dogs, childproof caps, and, of course, those damn Buns of Steel videos. Any of these could contribute to the problem, and the truth is, it may be more than one problem, each with its own cause.

Perhaps it is useful to start with a more specific mystery. Among the most vexing of the new diseases is Crohn's disease. You probably know someone with Crohn's disease. It is characterized by a suite of problems associated with attacks by the immune system on the gut, an internal turf war in which the immune system always wins. These attacks cause abdominal pain, skin rashes, arthritis, and even, in some cases, odd symptoms including inflammation of the eye. In the disease's most severe forms, Crohn's sufferers face years of vomiting, weight loss, debilitating cramps, and intestinal blockages. In these cases, sufferers often quit work to sit at home and force themselves to eat. Existing treatments are only sometimes effective. When individuals are badly afflicted, lengths of the intestine and colon are surgically removed, which, while it can be necessary in the short term, makes things progressively worse in the long term. Crohn's is rotten and debilitating and, except in rare cases, never goes away. It is also, rather suddenly, common.

In the 1930s, Crohn's was so rare that it was mostly undetected. Then, between 1950 and the mid-1980s, its incidence began to rise. In Olmstead County, Minnesota, the number of cases of Crohn's was ten times higher in 1980 than in 1940. The incidence of the disease has also risen precipitously in Nottingham, England, and Copenhagen, Denmark, and nearly all of the other places in the developed world for which the data are good. Today, roughly 600,000 people have Crohn's disease in the United States—accounting for some unnoted cases, about one in every 500 people. Similar proportions of people in much of Europe, Australia, and the more developed countries in Asia are also affected. From the perspective of the number of cases, Crohn's is a global epidemic, or at least an epidemic of developed countries.

Other than its consequences—the fates of the afflicted—just two things about the disease were, until very recently, certain: it has a genetic component (though weakly and inconsistently so) and is more common in smokers. But neither of these factors causes Crohn's disease. The average Kenyan can smoke all she wants and even if her brother in the United States has Crohn's, she still stands almost no chance of "catching" it. The gene variant that seems to predispose some individuals to Crohn's, CARD15, is not a requirement, nor is smoking, which seems to make the disease worse rather than actually triggering its onset. Somehow, economic development and what we tend to think of as modernity—affluence, urbanization, wealth—are prerequisites. It is as though progress itself makes us sick. For many years, inhabitants of India and China were unaffected, but now as India and China have become more successful, or at least some Indians and Chinese have become more successful, Crohn's disease has shown up there as well.

It may seem unusual that such a common disease is still poorly understood. The truth is that the causes of most diseases that afflict humans are not yet understood. More than 400 diseases that commonly affect humans have been named, and the unnamed diseases undoubtedly number in the hundreds. Perhaps a dozen of the named diseases—polio, smallpox, and malaria among them—are relatively well understood, but the vast majority, those other hundreds, are not. Though we may know how to treat the symptoms or kill the offending pathogen (if there is one) of the less well-understood diseases, precisely what happens in diseased bodies is, more often than not, a kind of corporeal mystery. What these poorly understood diseases almost inevitably have in common, though, is a few scientists dedicated to them, scientists who wake up thinking they finally understand what in the body is going on. In the case of Crohn's, one of those researchers is Jean-Pierre Hugot.

Hugot, a researcher at the Hôpital Robert Debré in Paris, thinks that the bacteria that live in refrigerators are to blame. Some evidence supports his theory and no evidence really contradicts it, but all he has to date is evidence that refrigerator bacteria are frequently found at the crime scene, a necessary but insufficient piece of evidence.[3] A recent study found that having a refrigerator in the home is indeed correlated with one's chance of developing Crohn's. But the study also found that having a TV, car, or washing machine was correlated with the probability of developing Crohn's. Another study found that Crohn's disease is less common where tuberculosis is more common. It is also more common where it is colder and where days are shorter. But a correlation between two things is no guarantee that one causes the other. There needs to be some kind of causal link and a demonstration of that link; one needs to show that A leads to B. Hugot had the A and the B, but not the "leads to." And so, although the refrigerator bacteria are found with Crohn's, they could just as easily be bystanders as villains. If it is not refrigeration, what is it?

Some biologists suggested pollution, others toothpaste or sulfur intake. Perhaps the measles vaccine? Or maybe Crohn's is psychosomatic. Maybe people in developed countries have idle minds prone to hypochondria.[4] The pattern in the distribution of where Crohn's does and does not occur seemed, like similar patterns for type 2 diabetes or schizophrenia, to invite wild ideas.

Whether or not one believes Hugot's speculation, one thing he said was right. Some species were favored and others disfavored by modernity. Hugot had focused on the species that were favored. But is it possible that Crohn's and other diseases of modernity have more to do with which species were disfavored? This was what Joel Weinstock, a medical researcher now at Tufts University and formerly at the University of Iowa, began to wonder. It was 1995 and he was on a flight to his home in Iowa from a meeting at the headquarters of the Crohn's and Colitis Foundation of America in New York.[5] He had just completed editing a book on parasites of the liver and intestines and was writing a review article on inflammatory bowel disease—a kind of medical catchall of diseases that includes Crohn's and other diseases that result from the immune system's attacks on the gut. Reading these two sources together made him conscious of the ways that parasites can harm their hosts, but also the ways in which they can help them, if only to ensure their own survival. In this light, it occurred to him that there was one thing that the family in Mumbai and the family in Manhattan shared besides refrigerators, TVs, and idle time. They were both missing something, namely experience with those species we had shed in our trip to the modern world, in particular our intestinal parasites—our worms. The germ theory of disease is based on the idea that we get sick when new species invade our bodies. Weinstock was thinking the opposite. Maybe some diseases are caused by taking species away.

It does not take much affluence to be wealthy enough to avoid intestinal worms. All you really have to do is wear shoes and use an indoor toilet. In the 1930s and 1940s, nearly half of American children had worms, whether big, twisty worms like *Ascaris*, tapeworms, or more delicate beasts like the small whipworm (*Trichuris trichuria*). Now worms are all but a thing of the past in the United States. Nor is the United States unusual. The places where Crohn's was becoming common seemed to Weinstock like the places where intestinal worms were known to have become rare. What if the absence of intestinal parasites was causing Crohn's disease? At that moment, Weinstock's idea was like so many other theories (albeit a little more unruly) in that it was simply correlative. Granted, it might be true that there were fewer parasites where Crohn's was more common, but as we've already seen, there were also more TVs and refrigerators. Yet Weinstock, thousands of feet in the air, felt confident, at least initially, that his speculation was right.

Wild speculation can be important to science, particularly in the early stages of a new field, when nearly anything is possible. In the early days, it seems as though anyone can solve the problem, so everyone tries. This stage of science can go on for decades, if not longer. It is from the initial blossoming of wild ideas that the truth is to be winnowed. But even if one accepts the fact that wild ideas are useful, some stretch the limits of well-behaved science. For, as strange as the refrigerator hypothesis seems, it was traditional medicine. In essence, the refrigerator hypothesis was based on the idea that some new species was infecting us and doing us harm. Hugot thought it was the cold-tolerant bacteria causing the disease. Other researchers have suggested that any of twenty other bacteria might be to blame.

What had occurred to Weinstock was something entirely different. His idea began with the observation that as we moved to cities and modernity, our bodies had lost rather than gained something. He thought it was the absence of parasites rather than the presence of a particular assailant that was hurting us. Our bodies, he imagined, missed their worms so badly that they were destroying themselves, eating their guts out in longing. As he sat cramped in his airplane seat, everything about Crohn's seemed clearer. Blue-collar workers were less likely to get Crohn's than people who sat all day at desks. They were also more likely to work in the dirt and to get parasitic worms! Suddenly this and a dozen other observations made sense. Weinstock had barely left the East Coast, but intellectually he felt as though he had just traveled a thousand miles. Everyone around him grumbled about their seats, the smell on the plane, and the surliness of the flight attendants. Oblivious to all of this, Weinstock was content.

There was precedent for the idea that one species, such as a human, might miss another species, even one such as an intestinal worm that had done it harm. The precedent involved pronghorn. The story of the pronghorn is relevant to Crohn's, and may be an answer where Crohn's and many of our modern chronic diseases are the question.

Pronghorn (*Antilocapra americana*) are small, goat-sized animals. Though they are sometimes called antelope, they are not precisely antelope and not precisely deer either. They are unique. Their branch on the tree of life has been separate from other branches for much longer than humans have been separate from other primates. Once there were many kinds of pronghorn but now there is just a single lithe species. The pronghorn's body is tan on the back and white on the belly. It has a dark black nose and dark black two-pronged antlers. Compared to elk, to moose, or even to true antelopes, pronghorn are dervishes—light and muscular. A pronghorn can run a hundred kilometers an hour. One scientist chasing pronghorn in the short grass of Colorado watched a few individuals run three kilometers and then put on a burst of speed. With that burst, they outran him, even though he was pursuing from a plane traveling 72 km per hour.[6] Even after running fast and long, they can run faster and longer, not faster than a speeding bullet, but faster, yes, than a pursuing plane.

Once, tens of millions of pronghorn thrived from Canada to Mexico. Then came the guns and greed of westward expansion. Pronghorn, like bison, were killed for food and sport until there were just a few million and then a few hundred thousand and then, finally, just thousands, a rare mother left here and there in the grass. Eventually, slowly, those thousands begat more thousands until, helped by land conservation, the pronghorn began to rebound. Today, 10 to 12 million pronghorn are alive at any one moment, scattered among the grasslands that remain. There, they bow to the ground to feed and then, at the slightest provocation, they run.

Counting pronghorn is difficult, like counting crows or clouds. They are suddenly everywhere and then, just as suddenly, nowhere. In most of the places they live, they remain unstudied, nameless, and totally wild. But there exists a grassland in the National Bison Range of Montana where the pronghorn are well-known. There the grass grows until it is about halfway up their backs and then stops. It bends in the wind to reveal them, in groups, looking back with their big brown eyes. The National Bison Range is still wild enough that things can live, mate, and die without ever being noticed, but it is defined enough, a world unto itself, that a man and a woman might hope to watch a few animals live out their lives and in doing so learn broader truths. So it was that in 1981, the zoologist John Byers took it upon himself to be such a man, and his wife, Karen, would be such a woman. John and Karen moved from Chicago to Moscow, Idaho, where he would begin as a new professor. From Moscow, when summer came, they would migrate to the Bison Range in an RV named Bucky. Rough but well loved, Bucky would launch them into the next phase of their lives.[7]

As John and Karen made their way toward the grasslands,

the landscape opened up. It looked like any grassland, as open and tanned as the savannas of Africa. Driving into it had the feeling of coming home, to a place where things are right and meaningful. They drove into the green-gray fescue, sage, and wheatgrass, and the forest disappeared behind them, and along with it, their daily existence. The space was wide open and yet complex. John would later write of it as the "floor of the sky."[8] It would hold them up for the summer or maybe even their lives.

When the Byerses arrived, they found the pronghorn. They watched them run until, at each fuzzy margin of visibility, they disappeared. The Byerses' first task was to catch these animals. Each captured individual would be tagged and then studied for as long as it or the couple might persist, years to be certain, maybe longer. But the catching was not easy. The adults were too fast and the fawns were, at least at first, difficult to find. But John and Karen persevered. Eventually, they found a mother with her two fawns hidden among the bladelike leaves of grass. As John approached, the mother fled, but the fawns froze. John reached down and picked them up in his big hands. They would be measured, weighed, and then tagged. They were, in their smallness, like birds, as at home in the air as on the ground. John and Karen hoped to follow these two and the others they would soon catch. Each fawn's heart pounded against the cage of its ribs until, into the air, it was freed.

John and Karen Byers settled into the grasslands with the idea of planting observations about the movements of pronghorn, their diet, mating, and everything else. Like any scientists, they hoped to watch one thing in order to understand others. They wanted to look to the pronghorn for broader truths. The pronghorn leaped and ran, and John and Karen saw in their running every living thing that runs. John and Karen lifted up the bodies of animals that they had caught and felt an example of any animal body.

Yet for as much as John and Karen thought they might find universal stories among the pronghorn, they kept finding ways in which the pronghorn seemed exceptional rather than general. One exception in particular had already plagued other scientists who had studied the pronghorn, or even just seen them—their speed. Audubon noted it, but so did everyone else who watched them for more than a few minutes. The pronghorn are faster at medium distances than cheetahs. They are twice as fast as wolves and faster than a safely driven camper van. It turns out they are even faster than an unsafely driven camper van. At medium distances, they may well be the fastest animal ever to have lived. That speed comes not from any particular biochemical magic, but instead from long thin legs, tiny near-featureless feet, an abundance of fast-twitch muscles, and hard-charging lungs. The pronghorn invest in speed at the expense of making bigger bodies or producing more young. They seem overbuilt, as though they evolved speed simply because it was possible. One scientific publication after another remarked on the pronghorn's speed.[9] Each one concluded it was anomalous, interesting, and just a little bit strange. Nor do the pronghorn simply run alone. They run and flee in tight groups more like schools of fish or flocks of birds than like anything on land, groups that move synchronously, their many legs in step, at high speed. The big question, aside from how, was why?

According to Darwin's rules, evolution does not overdesign. Natural selection is scrupulous in its editing. No material is wasted and no animal is taller, faster, or stronger than it needs to be in order to do better than its competitors. Were all the animals on Earth tortoises, there would be no advantage to being a hare, just the fastest tortoise. Yet the pronghorn, in their sinuous groups, outrun everything. In the thousands of hours the Byerses, other researchers, hunters, and locals have observed pronghorn, there are few recorded instances of adults being caught by predators. This is true even though many adult pronghorn have been dressed in radio collars and followed out across the plains and even though predation on fawns can be easily observed.[10] The fawns are eaten by eagles, coyotes, and other predators. But the fawns do not run in defense. They freeze. The adults are the ones that run and when they do, bears do not come close to catching them, nor do gray wolves or even coyotes. When the Byerses first saw the pronghorn's speed, it seemed like an affront to natural selection, a kind of gaudy exception flaunted at each opportunity.

John Byers was thinking about this exceptional speed when he started seeing phantoms. He saw animals chasing the pronghorn, sprinting after them. They caught them by their ankles. They took them down, one by one in among the lolling seed heads of the tallest grasses. They were not real, he knew that, but he could see their evidence, the way one might notice wind by seeing what it moves. In a landscape where the biggest predator is a bear, Byers saw the spoor of cheetahs and lions. If he squinted when he watched the pronghorn, he could even see these predators giving chase. He could see them manifest in the pronghorn's every action. Byers came to believe these ghosts were an answer to the question of the pronghorn's speed, and to other questions too.

As recently as 10,000 years ago, just as cows were beginning to be domesticated in Asia, the pronghorn lived on the plains with the gray wolf, black bear, grizzly bear, and coyote, but also with other large predators. When humans first arrived in the Americas, they found the pronghorn and alongside them a much greater variety of other herbivores, but, in addition, an even greater variety of predators. The American grasslands were more wild and ferocious than the African plains. The predators that the earliest immigrants to North America found as they colonized the continent fourteen or more thousand years ago were bigger, badder, and faster than anything we know today. There were plundering dogs (*Borophagus* spp.), short-legged dogs (*Protocyon* spp.), dire wolves (*Canus dirus*), giant cheetahs, giant cave lions (*Panthera atrox*), several kinds of saber-toothed cats, giant short-faced bears (*Arctodus simus*), and other toothy monsters, many of them fast. The cave lion grew to twelve feet in length. The saber-toothed cat could weigh 1,000 pounds and the giant short-faced bear as much as 2,500 pounds. Most relevant to the story of the pronghorn, though, was the American cheetah

(*Miracinonyx trumani*), a big, long, fast cat built to chase and catch at high speeds.[11] Analogies with modern African cheetahs suggest that the American cheetah would have loved to eat pronghorn the way African cheetahs love antelope. And so it was in this context that Byers began to imagine that the pronghorn's speed and its swarm-running evolved in response to now-extinct predators. The pronghorn once had something to flee. American cheetahs evolved to be faster to pursue faster pronghorn and pronghorn evolved to be even faster in return. Then humans arrived in the Americas and, one way or another, killed off sixty large mammal species, including the cheetah, but also lions, mammoths, mastodons, and even camels. The extinction of these great beasts and, in particular, of the American cheetah left the pronghorn anachronistically and irrelevantly fast.

Once Byers had this insight (which seems correct, in the hindsight provided by more data and analysis), much of the pronghorn's life seemed to make more sense. All of their biology, but particularly that of females, was built around escaping predators that were no longer present. Females chose fast males so that their children would stand a chance of being fast enough to escape. Even their double-horned uterus and compressed spine seemed a function of their past. They were not an exception, but instead a powerful manifestation of natural selection's rules. They were, in a way, the rule. What is more, it seems as though the pronghorn's speed and the traits related to it may be costly. If they are, and if, as time passes, the pronghorn become more abundant or their habitat more rare, the pronghorn may get slower. The individuals that run the fastest may die younger, exhausted from fleeing ghosts but unable to slow down. Given time, each generation of pronghorn might become slower and, in its more ordinary speed, less extraordinary.[12]

Here was what scientists all search for, a general result derived from the study of something very specific. Because the more Byers talked to other scientists, the more he realized that his case of pronghorn was not unique. His was one well-studied example of the consequences that result when one species misses another with which it had been linked for millennia. Years earlier, in Costa Rica, the tropical biologist and conservationist Dan Janzen had argued that the biggest fruits, those that now sit unmoved beneath their shady mothers, evolved to be dispersed by the now-extinct megafauna, species that disappeared along with the pronghorn's predators. Janzen's idea arose from his observations in 1979 of the three-foot-long pods of the *Cassia grandis* tree. Thirty years later, Janzen seems just as right and those fruits remain just as unmoved. To paraphrase the paleontologist Paul S. Martin, we live in a time of ghosts, their prehistoric presence hinted at by the largest sweet-tasting fruits.[13] Many of the fruits that humans have come to favor seem to have evolved to be carried from one place to another in the temporary vehicle of a giant mammal's guts—papayas make the list, as do avocados, guava, cherimoya, osage oranges, and the foul-smelling but delicious durian.[14] Elsewhere, biologists found long flowers without obvious pollinators, flowers that had evolved, they argued, in response to the long tongue of a now-extinct pollinator. With time, more cases like these have been noted, more examples of the consequences of losing partners middance.

But the pronghorn example was different. Giant fruits once benefited from being dispersed by giant fruit-eating mammals, sloths bigger than elephants and their kin. The pronghorn did not benefit from being eaten by a giant cheetah any more than you might benefit from getting eaten by, say, a bear. Yet without the cheetah, the pronghorn's lifestyle, its leaps and sprints, no longer makes complete sense. The pronghorn suffered from the American cheetahs that were their predators, but in a way the pronghorn may now suffer without their longtime foe present to give chase. They run for no reason. They waste energy, when they might do just as well to stand still. They run from ghosts.

We all do.

3

The Pronghorn Principle
and What Our Guts Flee

The Byerses went to the pronghorn to understand the pronghorn. What they found was more general. Let's call it the pronghorn principle. The pronghorn principle has two elements: First, all species have physical characteristics and genes that relate to the ways in which they interact with other species. Second, when those other species are removed, such features become anachronistic or worse. Plants have evolved toxins to defend their leaves, nectar to entice animals to carry their pollen, and fruits to attract other animals to carry their seeds. Animals, in turn, evolved long tongues to reach nectar or better senses of smell to detect fruit. Carnivores have long, sharp teeth to kill their prey. Intestinal parasites have appendages that mirror, in their contours, the guts of their hosts, to hang on. Pick any organism on Earth and as much of its biology is defined by how it interacts with other species as is influenced by the basics of living, eating, breathing, and mating. Interactions among species (what ecologists call interspecific interactions) are part of the tangled bank to which Darwin referred. What the Byerses newly understood in the context of the pronghorn was the consequences of removing the species our bodies evolved to interact with, be they predators (as in the case of the cheetah), mutualists like the animals that once dispersed the giant American fruits, or even parasites and disease. The loss of other species can make key elements of any organism's body as anachronistic as the giant fruits left sitting in the dirt, waiting for the megafauna that never come to pick them up.

For as much as the pronghorn principle is intuitive to ecologists and evolutionary biologists, a core part of our understanding of the living world, no one, not even the Byerses, had thought about how any of this related to our own bodies. Medical researchers are not usually trained to think about evolutionary history, and even when they are, they tend to learn about humans in isolation, as though the past was one of long, naked walks among the trees picking fruits to eat (yet even the fruit is another species, a species we had to see or smell to find). Until very recently, no research considered what happened when we killed off all of our predators, or, for that matter, removed the tapeworms, hookworms, and their kin from our guts. One wonders which parts of our bodies are, like the pronghorn's muscles and speed, haunted by ghosts—at least Joel Weinstock would come to wonder. What happens when humans leave behind the species their bodies evolved to interact with, whether they be cheetahs, diseases, honeybees, or giant sucking worms?

Joel Weinstock did not know anything about pronghorn. It is hard to imagine a circumstance in which they would have seemed relevant to him. Like most medical researchers, he had not taken anything resembling a class on ecology or evolution since he was eighteen. He could not have told you who the most recent ancestors of humans were, nor was he particularly a fan of "nature." He knew about the human immune system and how parasites affect it. These might seem narrow realms of biology to focus on, but in knowing two such fields, he was already broader than most biologists. This modest breadth became useful to him when flying home from his trip to New York. He flipped through a folder of data on the increasing frequency of Crohn's and other "modern" diseases and wondered why they had become more common. As he did, he remembered that over the same years, many of our parasitic worms had become rare. He connected these observations like dots. Once connected, they were a revelation. The cause of Crohn's could be, he suddenly thought, the parasitic worms—helminths! The more he looked at the dots he had connected, the more he thought he had the answer, an answer related more to the pronghorn and extinct cheetahs than to standard medical science.

It is joyful to think you have the answer. Your heart pounds. Maybe you run around the lab a little bit and let out a jungle yell. Of course, at some point you have to tell someone else your idea and that, in my experience, is where the raw insight often falls victim to reality. Some overly bright student says something like, "I don't understand how this would actually work," which is when you realize that it could not, and you sulk a little. But sometimes the insights are right. Or at least they seem right for a little longer than a single anticlimactic day.

Time would tell for Joel Weinstock. He imagined that the problem with our modern guts was our immune system, and that the problem with our immune system was that it was missing the parasites with which it had evolved. Crohn's and other inflammatory bowel diseases, he would come to argue, are the consequence of our body still running to escape its ancient assailant. When a pronghorn runs fast to outpace a long-gone predator, it wastes energy. When our bodies run fast to escape nonexistent worms, they trip, he believed, or maybe they never learn to run properly in the first place.

Weinstock had a hunch, but he did not have direct evidence. Of course, it was true that people in developed countries were more likely than those in developing countries to have Crohn's and less likely to have parasitic worms. In developing countries, as many as a billion people are infected with two species of hookworm alone (*Necator americanus* and *Ancylostoma duodenale*), not to mention tapeworms, whipworms, and other possible beasties one meets, however accidentally. All of these species were ancestrally marine creatures. They were able to come ashore by colonizing the guts of animals, each gut being, in a way, like a tiny, albeit not terribly picturesque, sea.

Such a possibility seemed crazy to the medical community, anathema to the long-held idea that medicine is meant, in part, to remove species from our bodies in order to make us healthy. Antibiotics, antiseptics, antihelminthics, and all of the other "antis" are based on the idea of removing life, regardless of its identity. Yet there was something to Weinstock's argument, and so people listened when he talked. He was also already, it bears mentioning, an esteemed immunologist when he started talking about "our bodies missing their tapeworms." Esteemed scientists suffer fewer consequences when they advance crazy ideas. They can shout them into a McDonald's takeout window. They can even announce them on TV. All that is likely to happen is that someone may say, "Damn, Joel, could you just test some of this stuff before you go on *Oprah?*"

Experiments are the best tests of new theories, but experiments on humans are not always possible, even when moral. It is hard to imagine an experiment to test the effects of refrigeration on Crohn's or any other disease. Refrigeration could play a role, but conclusive support is probably hard to ever come by. Even sick patients are unlikely to be willing to give up their refrigerators. The effects of the loss of parasites on Crohn's disease could, though, be tested experimentally. You would test it the same way you might test the effects of losing the megafauna on the American plains and the pronghorn. Reintroduce them. Restore the slender cheetahs of the gut, with their long tails and microscopic claws.

If losing parasites causes Crohn's disease, then putting them back might remedy Crohn's. But perhaps this is an overly simple idea, akin in that regard to rewilding the West to keep the pronghorn fast. If the experiment did not work, it would not tell you much. It might be that the effects of the loss of parasites are felt mostly due to their absence during immune system development, or to the loss of chronic infection. Might be. Could be. May be. Yet if you simply added parasites to a patient with Crohn's and the patient got better, it would be suggestive. If you added parasites to a bunch of patients and most of them got better, it would be more than suggestive, compelling even.

As I began to read the Crohn's disease literature, I wondered about such an experiment. Would it be moral? Would anyone approve it? Perhaps the clearest precedent for what Weinstock wanted to do came again from the pronghorn. A handful of scientists, friends of the pronghorn biologist Byers, suggests that we ought to rewild western North America. These scientists propose to "change the underlying premise of conservation biology." We need to, they say, reintroduce the extant carnivores everywhere they once were (bears and wolves, for example, occupy just one percent of the area they roamed just 200 years ago). But we also need to introduce elephants to replace mammoths and mastodons, African cheetahs to replace American cheetahs, and African lions to replace the extinct American lions. We might even introduce the Bactrian camel to fill the role of the many camel species once found in North America. By introducing these species to the American West, we would make it more like what it once was, what it, in their language, "should be." We would "stomp out the rats, dandelions and weeds." Maybe then, when the pronghorn again have something to flee, their speed will make sense.

These folks, with Josh Donlan, a conservation biologist currently based at Cornell University, as a kind of radical statesman leader, are macho mammal-trapping, snake-chasing dudes. They are ready for megafauna, ready for it now and unafraid of the consequences of advocating their new vision. They are the kinds of guys (and they are mostly guys) who, given their choice, would rather die in a tiger's mouth than of a heart attack. In one article Donlan asks, "Will you settle for an American wilderness emptier than it was just a hundred centuries ago?" Donlan will not. Bring back, he argues, the tigers. Bring back the lions. Donlan and his colleagues want these species back so badly that they are willing to go out into the desert and do it themselves. In fact that is just what they did.

Under the dark of night they caught a few wild animals from a preserve in Mexico, transported them in the back of a big truck across the border into Texas, and released them, wild and untethered onto Ted Turner's ranch. That the individuals were 100-pound Bolson tortoises (*Gopherus flavomarginatus*), not lions, and that the preserve was a fenced, albeit enormous, backyard is beside the point. The aim was, just as for the lions, to restore their functional roles. For the lions, of course, they would just need a bigger truck.

When Josh Donlan and others proposed rewilding the West, they got hate mail, or at least the academic version thereof, passive-aggressive papers written in response to their papers. The idea was taboo.[1] Then they got hate mail from farmers, whose predecessors and ancestors worked so very hard to get rid of the megafauna. In part, the critics' sentiment was an old one, summed up by the comments of the British biologist William Hunter some 240 years earlier when he wrote, "Though we may as philosophers regret it, as men we cannot but thank Heaven that its whole generation is probably extinct."[2] In other words, tigers are good in Bangladesh, but not in my backyard. Yet there is a key difference between the rewilding of the western United States, whether with a tortoise or a tiger, and the rewilding of our bodies. It is easier to get permission to do an experimental rewilding of a human body than of Idaho's wild, rippling miles of grass.

Donlan and other advocates of rewilding are still waiting for permission to release elephants and cheetahs in the Great Plains. They have had a few more successes, albeit not with mammals. The Aldabran tortoise was introduced, this time by the Danish ecologist Dennis Hansen, to a penned area of the island of Mauritius, where another species of giant tortoise once lived. Hansen has found some evidence that the reintroduced tortoises may help to restore populations of native plants by dispersing their seeds. Seedlings from seeds pooped by the tortoises grow taller, have more leaves, and are less likely to be eaten than those that simply fall to the ground. Whether any penned tortoises will be released is uncertain.[3] In the meantime, Weinstock and his colleagues started with an experiment on the habitats inside mice. They found that when you give mice nematode worms, you can prevent them from getting a mouse version of inflammatory bowel disease. With the mice as a wind at their sails, Joel Weinstock and his colleagues applied for permission from the University of Iowa Institutional Review Board to experimentally give human patients pig nematodes. Perhaps somewhat to their own surprise, they were approved.

In early 1999 patients with Crohn's were brought, one by one, into a lab in Iowa. They were given a survey and a medical exam to see if they met the requirements to participate in the experiment. Some were too sick. Some were pregnant. Some were too well. In the end, twenty-nine individuals were selected for the study. They were advised of the health risks of the experiment, which were largely unknown. They all agreed to have a radical theory tested in their bodies. If Weinstock was right, they might get well. If he was wrong, they would stay sick or potentially get even sicker. Either way they were to become host, if only briefly, to worms of a species kin to those we have spent millions of dollars eradicating. The progress of man was, in their bodies, about to be reversed.

When you are well, the body can feel invisible. When you are sick, the physicality of the body and all of its organs and tissues becomes all too clear, even exaggerated. Crohn's sufferers are reminded daily of the diverse ways in which the body, and digestion in particular, can fail. When things are working, we chew our food to break it down. We grind using our ancient teeth, teeth that arose in fish. Our tongue pushes the food down and our mouth coats it with saliva, which itself has enzymes such as amylase that help break down food. The resulting ball of slimy mash then passes through the stomach, where it is dissolved by acid and then on to the meters of the intestines, where each useful bit is absorbed into our bloodstream and from there trafficked on to our hot-firing cells. Amazingly, all of this machinery works most of the time for most of us. It works more often, in all likelihood, than any other machine you own, such as your garbage disposal or your car engine. But not for Crohn's patients. Crohn's patients are reminded, hourly, daily, often for their entire lives, of the limits of the body. They are reminded of the gut's vulgar occupation and weaknesses. At least some of them are reminded so vehemently that taking a treatment of pig worms seemed no stranger than their more daily experience of the body's uncomfortable failings.

While Weinstock's patients were prepared for the treatments, so were the pig nematodes (whipworms, to be specific). Weinstock and colleagues had to ensure that the worms were not carrying diseases from the guts of their origin. The eggs were taken from ordinary pigs and given to germ-free pigs. Like any Thoroughbred, these special worms were then allowed to mate, in the privacy of their pigs.* Their eggs were then harvested and divided into small piles of 2,500 each. The eggs look like small brown footballs with knobs on each end. Inside each egg, a living fetal worm was curled, balled up as tightly as the patient's hopes.

On March 14, 1999, twenty-nine patients, each of them sick and tired and worried, were given a glass of Gatorade with whipworm eggs suspended in it. The scientists added charcoal to the Gatorade to make the eggs invisible. The patients drank while watched by a study coordinator whose job it was to make sure that no one spit out their slurry. Each participant was willing to give the treatment a chance.[4] They swallowed the drink willingly, wiped their mouths, and waited.

Each patient was observed carefully. The experiment had been done in a kind of minitrial the year before with six patients suffering with extreme cases of Crohn's.[5] The results of this new, larger study were unpredictable. It was hoped that the pig worms would not attach in the patients' guts for long. They had not in the preliminary test, but the possibility couldn't be ruled out. The worms could have had negative effects. The patients were aware of this.

---

*One imagines, perhaps, a little Marvin Gaye music playing in the background.

They could have also searched libraries for "whipworm" or "*Trichuris*," and would have found a gallery of terrible beasts. Whipworms are like thin, featureless snakes. Mother whipworms produce thousands of eggs per day, each of which is, to use the euphemism, "deposited" by their host into the soil. Once in the soil, the eggs, if they do anything at all, wait. They wait to be accidentally ingested by someone else. Their improbable lineage has gone on like this for millions of years, one accident at a time. Back in the gut, the eggs hatch. The young worms crawl to the sides and find the mucosa where they complete their development and, once they reach adulthood, mate, though it was hoped this would not happen in the patients. The worms would, Weinstock thought, never mature, but instead simply evoke the desired immune response in patients.

A week went by. Two weeks went by. Each patient struggled to decide whether or not he or she was doing better. Four patients dropped out. More time passed. By week seven, some of the patients were feeling a little better, but some patients would have been feeling better anyway. At week twelve, the patients came back to the lab to be examined. Here was the test of Weinstock's radical rewilding. Then the results came in. They were announced by the lab manager on the phone. Twenty-two of the twenty-five patients still in the study were doing better. By week twenty-four, the last week of the study, all but one patient was doing better and twenty-one patients were in remission. These individuals, who had all been sick, were better. Their bodies were healthier now that they had parasites.

There are two ways to react to Weinstock's finding. The first is excitement. The second is concern about just why this effect occurs. Weinstock rewilded human guts and cured sick patients who had previously had little hope of getting better.[6] Nor were these patients with mild cases of Crohn's. These were the individuals whose disease had become untreatable by other means. Weinstock's study was just the beginning. His success inspired others. It was not long before other researchers suggested that many or most, or perhaps all autoimmune and allergic diseases were the result of missing our parasites. Perhaps even depression was linked to the lack of worms, and some cancers too. On the basis of this broader conjecturing, more experiments followed. If anything, these follow-ups, each seemingly more outrageous and significant than the one before, have provided evidence that Weinstock's underlying argument is ever more sound. When treated with worms, people with inflammatory bowel disease get better. Diabetic mice return to normal blood glucose levels.[7] The progression of heart disease is slowed. Even the symptoms of multiple sclerosis improve.

The eradication of helminth parasites from the developed world has been heralded as one of the major public health success stories, a symbol of our control of nature. But in context of the work of Weinstock and others, the consequences of our "control" are far from clear. To do better again we must bring back some of the worms (not all, obviously—many species of worms do have truly bad effects), carefully, the way, once channeled, the Mississippi has been allowed to flow again down some of its old diversions. We often view ourselves as separate from nature, but here is the rub: Our cultures have changed. Our behaviors have changed. Our diets have changed. Our medicine has changed. But our bodies are the same, essentially unaltered from 6,000 generations ago, when going for a run meant chasing after a wounded animal or fleeing a healthy one, water was drank out of cupped hands, and the sky still cracked wide open to reveal millions of stars, white dots as unexplainable as existence itself. Our bodies remember who we are. They respond as they have long responded, unaware that anything has changed, as anachronistically as the pronghorn's running or the megafauna's fat fruits.

Yet, as Rick Bass wrote in the foreword to one of John Byers's books about the pronghorn, "Almost never does one discovery tie things up neatly; rather it illuminates more unexplored territory and more unexamined patterns, one answer giving birth in that manner to a hundred more questions." The first of those hundred questions was simple: Why? Knowing that our bodies seem, in some real way, to need tapeworms, whipworms, hookworms, or the like does not really answer the simplest question: Why? Take the worms out and we get sick. Put them back and we get better. We could just go on putting them back and feeling, if besieged, also better. But before we intentionally add back to our bodies what we have long thought of as an adversary, it seems worthwhile to know what on earth (or rather, "in body") is going on. Whatever it is, it is happening to you right now, unless you already have a worm.

With time, Weinstock came to believe that the immune system requires the presence of worms to develop. Without worms, the immune system is like a plant left to grow in zero gravity. Long ago, in the evolution of land plants, conquering gravity's consequences was a major step in the plants' transition from swamp to land. Thick cells and strong—even woody—stems all evolved as means to cope with gravity, as did systems of transporting sugars, water, and gases. Nearly every difference between a tree and a swamp weed is a consequence of the greater difficulties plants face in dealing with gravity on land. In the absence of gravity, a plant's roots and shoots grow in every direction, like Medusa's wild hair. In a similar way, our immune system without parasites also seems to be challenged to distinguish up from down.

You may think I'm being too metaphorical. But to explain the relationship between worms and the immune system, immunologists themselves tend to turn more to metaphor and analogy than to fact. When pressed, Weinstock and others have begun to say things like, "without parasites, the immune system is in disequilibrium," or "disharmony," or in a more candid moment, "out of whack." "Different" is how one immunologist characterized the immune system of people in developed countries. This is the language of uncertainty. No one knows quite what happens when we take our parasites away. Take all of our worms away entirely, and we seem to stand a greater chance of getting sick. Put some of them back and we get better, much of the time anyway.

Weinstock has an idea for a more specific answer. Others do too, but the differences among what various scientists think might be going on are often at odds and hard to reconcile. Nonetheless, for now, Weinstock's version, a version initially offered by the immunologist Graham Rook at Cambridge University, is reasonable. That does not mean it is right, but at least given what we know today, it is possible.

Here it helps to know a little more about the human immune system. The body is a country with two immunological armed forces. One fights one kind of foe, viruses and bacteria; the other deals with another kind of foe, nematodes and other larger parasites. They work together, although the more the body's energy is spent on one part of the immune system, the less there is available to be spent on the other. This is a crude, almost cartoonish explanation, but even this much we have known only since the early 1980s. I could give you all the names and details, the TH1s, the TH2s, and the other untranslatable words of the immunologists' lexicon, but they serve, in this case, only to give the appearance of understanding, where we still have relatively little. So for the moment, just remember the two armed forces on different fronts, battling the inevitable enemies at our doors.

These two elements of our immune system have been in place for more than 200 million years. Sharks have them. Squirrels have them. Fish have them. Even some insects seem to have elements of them. They all have them because across the long history of animal lineages, each generation was filled with parasites, as well as bacteria and viruses. Our parasites were the ether in which our bodies made sense. The presence of these hangers-on has long been as dependable as gravity. Then it happened, the great change. Humans began to live in buildings and use toilets, and everything, in the last few generations—a second on a day clock of life's history—changed.

For a long time, we understood the immune system, in its near totality, as comprising just two main kinds of defensive forces, one against bacteria and viruses, the other against larger parasites. But in just the last five years, a problem emerged in this story. We were missing something, another character. What, scientists wondered, happened when parasites become ensconced in the body? The immune system, it was known, eventually stopped attacking them, but why and how?

It turns out we had completely missed a key component of the immune system, the peacekeepers. When a parasite is ensconced and initial attempts to expel it are unsuccessful, what should the body do? It could fight forever. In some cases this does happen and when it does, the disease and the problems caused by the body's immune response almost inevitably outweigh the trouble caused by the worm itself. In this context, the body may be better off giving in to the reality that the worm is present and learning to tolerate it. The answer appears to be, again and again, that if the parasite survives initially, the body learns to tolerate it. A team of peacekeeper cells calls off the antiparasite armed forces. The peacekeepers balance the response. They reserve the body's energy to fight another day against a more beatable or virulent foe.

What Weinstock, Rook, and others think is that these newly discovered peacekeepers are, in a way, our historical solution, but also our modern problem. The peacekeepers, they imagine, get produced only when there is peace to keep. When there are no ensconced parasites, particularly early in development, the peacekeepers dither and wither. But the forces remain strong, and so without being otherwise occupied, they attack whatever seems foreign. They can sometimes be so eager to win that they fight whatever they come across. The body's own bits and pieces begin to seem threatening. The peacekeepers that might otherwise call these increasingly indiscriminate forces back, do not. They are too weak. Unchecked, our immune system battles our bodies without end. It battles our bodies until we are sick and then sicker. Boils erupt on our skin. Our intestines become inflamed. Our lungs wheeze and collapse. It battles our bodies until there are no winners.

Weinstock thinks that when he introduced worms into patients, their bodies began to produce peacekeepers, which kept the peace by stopping the immune system from attacking the worms. Of course, just like the cheetahs that pursue pronghorn, hookworms can have costs, the most common of which is the loss of blood in severe infections and consequent anemia. But on average the costs appear minimal, both in a general sense and relative to the costs of fighting the worm forever. If the worm is well ensconced and the body continues to fight it (or them), the body wastes energy. And so it may be that the peacekeepers provide a mechanism for the gut to admit local defeat and at the same time prevent the immune system from a prolonged attack on the gut, whether a worm is present or in other situations. The peacekeepers keep the peace. The worms, in their way, trigger that peace. Maybe.

A second possibility also exists and this possibility (which is not really exclusive of the first) is my own favorite. It has long been known that worms in our guts can produce compounds that suppress the immune system, compounds that, in essence, signal, "Hey, it's cool in here—no need to attack." They do so by mimicking some of the body's own compounds. Many different worms produce these compounds. It may be that our bodies evolved to depend on at least low levels of such worm-produced compounds. Here I do not mean that our bodies needed them, at least not originally, so much as that they could count on their always being there. Perhaps our bodies produce more of an immune response than is necessary because they are, in a way, "assuming" that some of their response will be dulled by the worms. No one can show that such a phenomenon is occurring, not yet, but it seems plausible.

In the meantime, the broader reality is that our immune systems appear to have evolved in such a way as to function "normally" only when worms are present. Scientists other than Weinstock have called this phenomenon the hygiene hypothesis, where the idea is that clean living is bad for us because the functioning of our immune systems needs the "dirty" realities of worms and maybe even a particular microbe or two. What seems to have been missed is that it is not just our immune system that evolved to depend on the presence of other species. It is the shape of our guts, the enzymes we produce in our mouths, and even our vision, brains, and culture too. All of these parts of our lives evolved with the gravity of other species as a forgone conclusion. Then we removed or changed those species and in doing so, altered the biological gravity of our lives. As in the case of worms, it is not always clear what it is about having cleaned them away that causes us trouble, but it does seem clear that again and again it changes us, leaves us like a ballroom dancer with her arms held in position but nobody to hold.

We will return to the parts of our bodies shaped by the influence of other species, be they mutualists in our guts, mutualists in our fields, predators, or pathogens. Meanwhile, as you sit and read, something is happening in your gut; the forces are arming themselves. Whether or not you have a worm will influence just what they do (as will, almost certainly, other factors, such as the particulars of your genes). Your immune system is acting on your behalf, but without conscious control. It is acting right now, with armies of tiny structures. If your immune system has not turned on you, with allergies, diabetes, Crohn's, or other problems, you have good luck, good genes, a good worm, or all of the above. But many of our immune systems will turn on us eventually. If yours does, the question is what you should do. Do you, could you, would you, search out a worm?

4

The Dirty Realities of What to Do When You Are Sick and Missing Your Worms

Whether or not you suffer from diseases related to your immune system, millions of other people do. The afflicted have only so much patience for the slow pace of science. Waiting for theory can be like waiting to die; both are deeply unsatisfying. Debora Wade was not one to wait. By 2006 or 2007, she had read up and knew about the missing worms and Joel Weinstock. She was sick of waiting and not knowing what was on the horizon, but mostly just sick of Crohn's disease. For twenty grinding years, she had walked around not knowing when she would have to run to the bathroom. She wanted to get better, but like nearly everyone else with Crohn's, she had few options. Her life had gone on, getting more and more besieged by the daily realities of Crohn's.

Debora reached the point where she felt like she was willing to do anything. She had read about Weinstock's glasses full of Gatorade and whipworms. The idea was disgusting, and yet somehow it was as appealing as any other option. The drugs she was taking for the Crohn's were not working. Neither was she. She could not. She was sick, housebound, and emaciated. She had "chronic diarrhea, night sweats, and painful bowels." She could sometimes digest blended soup, but not always. The flavors, it seemed, had been sucked out of her life.

She searched the Internet, went to the library, called friends, talked to researchers, and otherwise turned every stone. She had lost precise count of the exact number and kinds of pills she had taken, but she could remember their consequences. Her doctor told her about another new experimental treatment. This latest treatment offered, as if a kind of punishment for wanting to get better, a high risk of cancer. She started over, looking, asking, and doing what she could. Eventually she circled back to where she had started—the worms. The thought appalled her, but compared to the alternatives, untested and intensive chemotherapy, bone marrow transplants, or worse, how bad could it be? Being dosed with worms hardly seemed more barbaric than what she had already been through. She decided that she would do it, maybe. No, she would definitely do it.

In talking to her doctor, she committed herself to a dose of mail-order whipworm eggs, the same eggs Weinstock's technician had offered to patients in a glass of Gatorade. But when she went to order the eggs, she found a new problem. The U.S. Food and Drug Administration had made it illegal to ship whipworm eggs, and even if she could get the mail-order worms, they would cost $4,700 for the first two weeks. They then had to be taken again and again, she assumed, although for how long, no one actually knew or knows, to remind the body that the worms were there; $4,700 repeated monthly, perhaps over a lifetime.

Debora was, once more, out of options. Then she heard about a study in Nottingham, England, in which patients were being given whipworms in a double-blind experiment (some patients would receive the worms, others a placebo) to test their effects on allergic rhinitis, asthma, and Crohn's disease. She called. She tried to stay calm, but could not avoid being a little hopeful. The study was willing to take Americans as patients! She found herself feeling optimistic for the first time. Then she learned about the catch. She would have to visit the United Kingdom six times over the course of a year. In her condition, she could not imagine flying on a plane once, much less six times or more. She was too sick, and even if she paid her way, again, again, and again, she had a fifty-fifty chance of receiving a placebo.

Then she saw it, blinking like a small beacon of Internet marketing hope on her computer screen. You can search for it yourself. The last time I looked, it came up when I typed "treatment for Crohn's," "experimental treatment," and "help" into a search engine. For just $3,900 U.S., Debora Wade could travel to Mexico and have someone give her a dose of hookworms. These were not pig whipworms or even pig hookworms, but instead, good old-fashioned human hookworms, the hookworms that crawl through your skin and into your body, and in doing so can make you anemic or, very occasionally, worse. She found an advertisement offering to give those to you for the price of a used car. Was she losing her mind? This was not modern medicine—it was a guy hawking parasites from a clinic in Tijuana. It was not even clear that he had a medical degree (he did not). Yet what were her alternatives? She was not dying, but the life she had grown up imagining for herself seemed to have been.

She told her doctor about the Mexican clinic, and he told her not to go, though he understood her temptation. Modern medicine is built around the idea of holding back other species, but when the problem is that those species are missing, modern medicine throws up its hands and offers a shrug and another drug. So too did Wade's doctor. Yet for Wade, it was not just that the worms might make her better; it was that they were, in their way, easier than anything else she had been doing—the pills, the shots, and the constant tending of her body's reluctant garden. If everything worked, the hookworms would crawl through her arm into her bloodstream, where they would ride past the heart into the lungs, get coughed up from the lungs, swallowed, and then from there make a road trip down into her intestines. In the intestines, they would ensconce themselves and live out their lives, three to seven years or, if she was good to them, even longer. If she happened to get a male and female, they might breed, but they would not increase in number. Their eggs would be passed into her toilet and down into the sewer system of Santa Cruz, California, where she lives. She would not have to take a glass full of whipworm eggs every two weeks. She could just take the treatment once every three years—or maybe less than that, once every decade. The endeavor felt more like adopting a pet than modern medicine—a long, translucent, sucker-mouthed companion animal.

The "treatment" carried risks. Her family reminded her of them. Yet she knew everything she had already been doing had risks. The "cutting-edge" treatments caused cancer or infections. The cutting-edge treatments were just as poorly studied as the worms whose risks had already been weighed across the bodies of millions of people. Their effects, though sometimes rotten, were also predictable—at least it seemed so at the time. So she got in a car with her family and headed south on Route 5 toward Mexico.

As Debora Wade drove to Mexico, many things went through her head. Her doctor had warned her that in Mexico she would have no control over what she was being given. She had no guarantee, he argued, that they would even be hookworms. She had no guarantee where the hookworms had come from. Would the hookworms be clean of other parasites, viruses, and bacteria, for example? She did not and could not know, yet somehow she felt good. She felt as though she had made one of the most important decisions of her life, that for the first time in twenty years she finally, really, had a shot at feeling better.

The man Debora Wade was driving to see was Jasper Lawrence. She had never met him, but she knew his story—everyone seemed to. Once heard, it is unforgettable.

For many years of his life, Lawrence had worked at an advertising agency in Silicon Valley. He was successful, but also sick with asthma, an affliction that caused him to worry, prematurely, about his own death. Whenever he inhaled, his lungs felt fragile to him—one breath away from failing. He had always been sick, even as a child, but lately his health, particularly his asthma, had gotten worse. He was guilty of the usual sin of smoking, but whether this alone was why he was sick or whether he was the victim of some more complicated series of consequences and genes is unknowable. Regardless, though, of how his story is reconstructed, it led him to a point in his life when he was in and out of the hospital and entirely dependent on Prednisone steroid tablets to go about his daily life. Then, unrelated to his health problems, he took a new job. He was excited about the new position and the change it would bring to his life. Little did he know.

In the process of starting his new job, Lawrence had to start a new insurance plan. He needed insurance to pay for his medicine, but his preexisting condition ruled him out and he was left uninsured and scared. He was scared for his life. One can imagine his headstone: "Here lies Jasper Lawrence. He died of preexisting conditions." He breathed shallowly, conscious of each breath.

Lawrence was not yet in a crisis but he saw it ahead of him, looming. His problems made him open to possibilities he would have missed at other points in his life. He was prepared for big change and so when a kind of opportunity presented itself, he did not ignore it. "The change" began on a routine visit to his aunt in the United Kingdom. During the visit, he was up late one night, unable to sleep. Restless but unfocused, he sat down and began looking for solutions on a computer. His aunt had mentioned a BBC documentary about hookworms and diseases such as multiple sclerosis and asthma. As he began looking for the documentary, he found research articles by Joel Weinstock and other scientists. He started reading them, at first out of curiosity and then with a tingling excitement. He doesn't remember quite what he saw that day. Maybe there was a study about dosages, or how many worms one needed to have in one's gut to have an effect. What he does remember is that by the time he went to bed, with the sun already rising, he had decided that he was going "to try worms of some variety."

He would, he decided, do whatever it took. At worst, he thought he would look foolish. At best, he would be healed. All night he dreamed of worms, tangling, wriggling, and crawling. They were, at least in retrospect, good dreams.

Lawrence spent the next months reading and rereading whatever he could find about worms and health. He wasn't a scientist and so, particularly at first, the articles were difficult for him, as they are for most people who decide to take their health into their own hands. The science was hard to understand because of the scientists' jargon, but also because the scientists themselves didn't know exactly what was going on, what happened when the proverbial hammer hit the nail. Yet Lawrence, as he read, was ever surer that what he needed was worms. His first practical problems, like those of Debora Wade, were to choose and then get a worm. The science published so far is mute as to whether one kind of worm is any better than another. The options are nearly endless. Testicle-enlarging worms; whipworms; tapeworms, worms that grow to thirty feet in your colon. Choices, choices. Lawrence came to favor using hookworms. Tapeworms, he thought, were the most likely to actually cure him, but they brought the possibility for reinfection. Call him squeamish, but he simply didn't like the idea of a thirty-foot monster living in his gut, much less the possibility of "passing" such a beast.* His decisions were not quite educated but they would do.

Yet as much as he tried over the next eighteen months, Lawrence could not come up with a good way to get hookworms. At the same time, his confidence that the worms might heal him grew, as did his sense that he was getting sicker. He brooded. He made phone calls. He read more. Eventually, he came to terms with the realization that short of some treatment in a medical setting (which, at the time, did not exist), he needed to get hookworms

---

*One parasitologist I talked to referred to this process less euphemistically as "shitting a dragon." I guess that, technically, it is not a euphemism.

himself, the old-fashioned way, from someone else. He learned from reading that hookworms were nearly everywhere in much of the poor world and, much to the distress of public health officials, easy to get in those regions, at least when you did not want them. So he decided he would get a do-it-yourself worm. He bought a ticket to Cameroon. In Cameroon, hookworms infect nearly everyone. In Cameroon, all he would have to do was be a little bit incautious and he would get a hookworm. If a little bit incautious did not work, he was prepared to do more.

Lawrence flew to the United Kingdom and then from the United Kingdom to Cameroon. The trip was expensive and bold. Lawrence was a California native, an entrepreneur. He was not the kind of guy who traveled to developing countries, but here he was, arriving in an airport that looked to him more like an old high school than someplace to be charged with the safety of planes. He got out of the plane and the air was hot. He saw poverty everywhere. In the days that followed, he saw fingerless lepers, begging children, bus accidents, and a great and terrible disregard for life. Lawrence was also seeing the irony, though irony is not a strong enough word, of what he was doing. Much of the world, including Cameroon, remains unable to rid itself of the parasites that end lives prematurely and brutally. HIV, malaria, and dengue kill people, destabilize governments, and even precipitate wars. Alongside these other diseases, the worms too are thought to ruin lives. But Lawrence, like Joel Weinstock, whose work he had read, had come to think that the story about worms was more complicated. He had come, with the instrument of his body, to test that belief.

Lawrence stayed with a family he had met on the plane. He explained to them his plan to contract hookworms. They must have thought him mad, but he was hardly the first Westerner to travel half mad to the jungle in search of a cure or treasure. He was not so different from those earlier explorers except that, unlike the rest of them, he wanted to go to the poorest, dirtiest places in the country. He wanted to go there barefoot so that he might contract hookworms. Surely this wasn't the best way, but if he didn't get better, he would go broke. If he went broke and ended up without Prednisone, he would die. So he found himself in Cameroon, looking for piles of human excrement to walk through on the chance that in those piles a few worms might crawl into him, through the thin barrier of his soft, urban skin.

Actually, he did not have to walk through fresh excrement. Much to his initial dismay, it took Lawrence several wet, stinking-footed trips to latrines before he learned this important fact. Hookworms take days to mature and so he needed only to walk in older, dry latrines—the places "behind the house" where narrow footpaths end in a hole in the ground or more often a series of piles of toilet paper and shit. As he looked for these places, people yelled at him. They chased him. He defended himself with what, given the half lunacy he was engaged in, stood for reason. But the more he explained, the angrier people seemed to become. The only way this story seems as if it could end is with a fistfight or a stick fight or worse between him and some man out in the excrement piles of Cameroon, the grass up to their waists, hookworms crawling around the ground, a desperate, bloody, mad fight for one man's survival and the other's dignity. Somehow it did not. Somehow Lawrence went from latrine to latrine without getting bludgeoned until, one day, he felt an itch on his foot. It was the itch of good luck, the itch of worms crawling through his thin, once-affluent skin, headed, he could only hope, for his heart.

When Debora Wade decided to go to Mexico, she knew Lawrence had gone to Cameroon to seek out his own treatment. She probably also knew the lengths to which Lawrence went to get his worms. Most important, she knew the punch line to his story: that the worms had crawled into his bloodstream, gone through his heart, and made it to his intestines. She knew too that once there, they had somehow engaged his immune system and that the end result was, in some measurable way, the near complete disappearance of his asthma. His immune system no longer attacked pollen. His immune system no longer responded to every allergen. His immune system, in fact, no longer triggered any of the events that had so plagued him. He breathed easily. So seemingly miraculous were his results that he had decided it was his life's work to help other people seek treatment, not in Cameroon but somewhere nearer his home. He started his clinic in Mexico to distribute worms. Lawrence was a changed man and you could be too, or so the former ad-agency executive's Web site claimed.

Debora Wade was worried as she headed to Tijuana. She had never been to Mexico before because she had been afraid of getting a parasite. Her overwhelming thought was "What the hell am I doing?" On December 17, 2007, she had made it to San Diego, where she would cross over the border into her new parasitized life. She went with her family to a resort hotel (surely a little pampering is justified prior to intentional infection with worms) where they would relax, but even before she made it to her room, Jasper was there, at the front desk, to greet her, to shake her hand and introduce himself. Tomorrow, the infection would begin.

At the hotel, Wade slept well, but then awoke nervous, her heart pounding in her throat and adrenaline overwhelming her body. She got in a car and drove to the clinic. The neighborhood looked rough. The clinic itself was a two-story house on a busy street, Lawrence's house. Inside, Debora found Lawrence again, along with a man named Dr. Llamas, who would actually initiate the infection. She waited briefly beside her husband in a waiting room/living room before being ushered down the hallway to a room like you might find in any doctor's office. The room had a hospital bench like she would have seen at home, covered in white paper. Dr. Llamas was friendly. He asked about her illness. He asked about her health. He sympathized. As he did, a nurse came into the office to take some blood. It was in that moment that it really dawned on her how little control she had over what she was about to be given and what the consequences might be. She was at the mercy of Dr. Llamas and

Lawrence, but also of the behavior of an undomesticated group of worms, worms with no interest in her concerns or fate, wild worms. The next day the potential fireworks began. Llamas infected her with, she hoped, wellness. If things went well, the worm larvae— offspring of those larvae currently in Jasper Lawrence—would crawl into her skin. When the procedure was done, she thanked everyone. She collected her husband and headed home.

To date, almost a hundred patients with asthma, ulcerative colitis, Crohn's disease, and other autoimmune maladies have made the trip Debora made, to this same office. As for Debora Wade, once home, she waited. The whole procedure was done but it was, she had to conclude, less satisfying than she had imagined. For one thing, the travel to Mexico and the treatment itself cost nearly $8,000. For another, the worms, her worms, were donated by Lawrence. It had not initially occurred to her that they would come from someone. Jasper Lawrence passed the eggs that became worms that were eventually given to her. Did she really know Lawrence? Had she checked his credentials? As she replayed the entire scene in her head, she recalled that the man who had drawn blood had dirty nails. The office was, well, not entirely clean either. What had she done?

Once home, she peeled off the bandage, looking for a series of ten dots, points of origin where the larvae had burrowed in. She found a single red dot, but nothing else. The beginnings of this whole endeavor were, to put it lightly, anticlimactic. On the third day, she was very clearly still sick and spent the night on the toilet looking at constellations out the window. More time passed. Christmas came and, with it, a fever, perhaps in response to the worms settling in, her body fighting what her mind so badly wanted. Then she got sicker, both with Crohn's and now with a fever. She could scarcely consider the thought, but she felt as though she were getting worse. Finally, and even then only slowly, she seemed to be getting better. But it was so hard to tell, to measure progress against hope.

Then, very clearly, Debora got worse. New symptoms turned

up: arthritis and swollen ankles along with the old symptoms.[1] Then she got better, much better. For a while she was in total remission. She was, amazingly, no longer, for those days, a Crohn's patient. Then things got worse again. Ultimately, she was reinoculated with worms. It seems that after each inoculation, she feels better for a few months, and then symptoms return, creeping back, perhaps as her worms die. As of June 2010, she was preparing to go back again, to get more worms. She goes on living like this, day to day, rewilding her body and while not yet cured, better. That is all she had asked.

We want medicine to be sophisticated and effective and informed by knowledge. In the old days, both the Egyptians and the Incans augured holes in people's skulls to heal them.[2] Sometimes it worked. In other instances, the patient died with a drill stuck in his head. All surgeons have good days and bad days, but ideally we want to know what distinguishes the two, and to err in the direction of the former. Debora Wade's story makes clear that, as often as not, what we know about treatments is that they work—sometimes. We still view our bodies like machines, in need of a little hammering here, some welding there, and the occasional drop of some chemicals to clean us out. They are not machines. They are organisms that evolved in the context of other wild species, organisms full of particulars, organisms that, despite several centuries of medical science, remain fundamentally full of mystery. We need more information, but in particular we need more information about the evolutionary and ecological conditions that have led to a problem in the first place.

Our prevailing form of treatment for difficult illnesses like Crohn's is to use medicines that treat the symptoms—but this is, at best, a thumb in the levee. Consequently, when it comes to the question of whether or not one should get treated with worms when one has Crohn's or diabetes or something else, we are still in the Wild West. We know so little that the medical community can't really provide a good answer. The worms, it is clear, are not a simple panacea. The worms don't seem to work for everyone. Jasper Lawrence estimated that about two thirds of his treatments are successful, though with imperfect follow-up (some patients simply go home and disappear), it is hard to know. Debora Wade thinks that 70 percent or so of the patients she has talked to seem to have improved. There are, among those patients, miracle stories. Two multiple sclerosis patients have been in total remission for two years. Many of the allergy and asthma sufferers seem to be healed. On the other hand, other individuals, some with the same diseases, have had less success. Patients with ulcerative colitis seem to have had little luck with the treatment. Debora Wade is in contact with several of the Crohn's patients. Three of the patients, like Debora, initially felt much better, but after six months the effect had worn off, perhaps because the worms died. Reinoculation seems to help.

Debora Wade, though she has had more ambiguous results than many who have been treated, still swears by her worms. She goes on reinfecting herself. Most days she feels better. She now has new symptoms, the cause of which she can't discern, but the same thing would have been true had she tried the new pill or injection or whatever chemical was about to come next. Meanwhile, she and others wait for more research. As Debora told me, "It is all very new and we have no idea what we are doing, if more is better," no idea what the required frequency of reinfection is, or anything else. Other research is ongoing, though from Debora's perspective, too slowly. The Nottingham study in which she first thought to enroll has finished. They have not yet published their results.

Dr. David Pritchard, the biologist in charge of that study, is moving forward with trepidation. The fact that so many people are being treated before the treatment is well understood is worrisome to Pritchard. Yet so few people work on the effects of helminths or other parasites and diseases on the immune system, particularly in a clinical setting, that patients who take matters into their own hands might also be doing what is reasonable. Outside of Lawrence's treatments and the experiments in Nottingham, there is ongoing research in Edinburgh and London, work by Weinstock in the United States and a new project in Australia. There are two more sites in Mexico where worm treatments are being done, at Ovamed and Wormtherapy, the latter run by Garin Aglietti, one of Jasper Lawrence's former collaborators who broke ranks.

In a way, Dr. Pritchard in Nottingham is undoubtedly right: What is happening south of the border, in Mexico, is wild. What Jasper Lawrence is doing is unproven but is not an experiment in the scientific sense. In other words, there is no control, no real monitoring of results, and no comparison to what happens to patients that go untreated.

So if you have Crohn's, what should you do? If you have allergies or diabetes or inflammatory bowel syndrome or MS and are desperate for a healthier life, is there hope? It seems clear that parasites and these diseases are related, but it is less clear how they are related. It seems we need, somehow, to get back to some version of the old days, but the old days are gone, and we need to come up with a new way of restoring elements of what once was. We need, in a way, to domesticate our worms, to make their effects more predictable and their consequences more controlled. For now, for those who are suffering versions of the diseases related to the loss of worms, diseases that are recalcitrant to standard treatments, there are few choices. What would I do in that situation? I would probably travel barefoot to one of those regions where worms are the norm, but I would choose carefully. Or maybe I'm lucky; I've traveled enough and walked barefoot enough, so that I might already have some. There are no perfect options. These are the dirty realities of our situation, where we remain bound to our history in webs so complicated that we can't quite untangle them, not yet anyway.

The lesson that the worms clearly offer, though, is that the old medical model, in which we just scrub the rest of life off our bodies, is wrong. Major systems of our bodies, including our immune system, evolved to work best when other species lived on us. We are not simply hosts to other species; we live lives intimately linked to them, and even the boundaries between the simplest categories of "us" and "them" and "good" and "bad" are blurry to the tools we have so far. And the worms are just the beginning. On our bodies are thousands of species, a kind of living wonderland. There are more bacterial cells on you right now than there ever were bison on the Great Plains, more microbial cells, in fact, than human cells. It is to the question of those cells, each one of them tiny but perhaps consequential, and their relationship to our well-being, that we turn now. No human is an island, not even when she is free of worms.

PART III

What Your Appendix Does and How It Has Changed

5

Several Things the Gut Knows and the Brain Ignores

Once we learn how to kill something, we tend to do so. We enjoy the hunt. With stone-tipped spears, we stabbed at mastodons. We chased saber-toothed tigers, dire wolves, and the American cheetahs that once ate pronghorn. There was a rush in the pursuit. With guns, we did the same job even more exhaustively, until eventually we moved on to smaller prey such as the passenger pigeon, animals that we would sometimes eat, but more often not. The urge to hunt can be greater than our needs. After the invention of pesticides, we sprayed millions of acres for smaller quarry. We even sprayed our bodies. DDT was rubbed lovingly into the hair of hundreds of thousands of children. Once we learned to harness compounds that would kill microbes, we filled ourselves with these concoctions. As much as we might love landscape paintings and fleeting glimpses of wildlife, nothing seems more natural to our brains than getting rid of nature.

Each of the technologies we have used against other species is a kind of anti-biotic (literally, "against life")—though seldom does a technology actually kill all of the life we are after. Instead, each tends to favor some over others, the strong or weedy over the weak and slow growing. When we stoned, speared, or shot big predators, smaller predators did better.[1] We used DDT to kill the pests on our crops and in our homes, and favored the resistant and insidious. We sprayed our crops and yards to kill the weeds and left the super-weeds to grow up between our rows of corn and out of the cracks in our cement. All around us we find these species, like dandelions and ragweed, species that blossom out of hardship and persistence, growing toward the sun even as they shake the asphalt from their leaves.

If the pronghorn is an experiment on the effects of removing a predator from its prey, we are the broader experiment. We are a case study in the effects of removing not only predators but also snakes, intestinal worms, and even microbes, to see what happens and who remains. The extent of this experiment is greatest at its most intimate, in and on our bodies. We have removed our worms, but more recently we have also begun to remove, or try to remove, our bacteria and other single-celled life-forms, this time with antimicrobial agents. These agents are what we most often think of when we say "antibiotic," the compounds originally produced by fungi like bread mold and discovered by Alexander Fleming, the haphazard visionary of life. It would be fair to wonder which species they kill and which they favor. After all, you have probably used antibiotics. Even if you have not intentionally used them, you have ingested them. They are in our food and drinks. They are used on crops, in cows, pigs, and other domestic animals both to treat bacterial diseases and to prevent their occurring in the first place. Antibiotics are nearly everywhere. More than 200,000 tons of antibiotics are consumed annually,[2] with more consumed both per person and overall each year. Scrub. Wash your hands. Scrub again. Kill what grows before it spreads and then kill it again. This is what we long have done, what our ancestors did, and, without vision or change, what we will do in the future. It is what comes naturally.

We began using antibiotics because we needed them, desperately. Their discovery yielded a trio of Nobel Prizes and cured our gonorrhea, tuberculosis, and syphilis in the process.[3] Penicillin was the most effective life-saving drug in the history of the world, rivaled only by other antibiotics. But the use of antibiotics for the treatment of deadly diseases now represents a tiny proportion of all uses—most are for sniffles, earaches, or even preemptive attempts to ward off microbial evil. ("Doctor, I'm feeling a little funny. I think I might be getting, well, I don't know, something that needs antibiotics . . .") or so the story, again and again, goes. We turn easily to pills or spoonfuls of amoxicillin, ampicillin, good ole penicillin and all the rest. We turn to them as we once turned to our guns, in self-defense. The question is not whether our antibiotics, in the most general use of the term, have helped us, but instead, when we pull the trigger, how well we are able to aim.

For most of the long history of antibiotics, no one studied the details of how they affected the bacteria in our guts. The approach of medical research is often to see what helps us first and then, only secondarily, to understand how and why it works. It was known that antibiotics kill pathogens such as syphilis (we know that because when patients are given the antibiotic, the syphilis goes away). But what actually happened to the other microbes in and on us when the syphilis was dying was never studied. The appropriate technology did not exist. And, more to the point, for the medical research community the goal was curing diseases. Many diseases were bacterial in origin and so all bacteria came to be considered bad (an idea perpetuated by James Reyniers, the bubble rat king, to whom we will return). They were as bad as the leopards and wolves that once ate our animals and children or as the pests that consumed our crops and sustenance. "Kill them all now and ask questions later" was the medical solution. At least initially, this approach seemed reasonable.

I understand our urge to go a little rogue when we first invent a new tool, particularly when it is in the interest of survival. When some kind of life is discovered to be both controllable and at the root of some ill, we control it. Yet at the same time, when we learn to distinguish the riff from the raff, the deadly from the innocuous or even beneficial, we also ought to try to kill with nuance. The problem in our guts is that until very recently we could not distinguish riff from raff, nor did we even know which species our weapons—in this case antibiotics—were killing. Or perhaps I should say that our brains could not make these distinctions, because it would turn out that our guts (and in particular our appendices) knew what was going on the whole time. They just couldn't say anything.

The reasons for our ignorance of the goings on inside our bodies are straightforward. Our guts may be as unknown as tropical forest canopies, but they lack in both scenic beauty and romance. If you work in the rainforest, people you bump into at dinner parties will mention their plans to someday travel to Brazil or Costa Rica. Work in the colon and people will mention, at best, their lunch. At worst, well, you can imagine. . . . But it is not just that the gut is unsexy. It is also difficult. The species that live in the rainforest canopy can be taken back to the lab or the field station to be observed, poked, and prodded. We can see what they eat and even watch how they behave. Not so for the gut microbes, most of which are both invisible and unculturable. More than a thousand species of microbes have been found in human guts. A thousand more may live on your other parts. Most of them we cannot grow at all, except where we find them. We know too little about them to get them to live in the lab. They are alive and in us, yet inscrutably difficult to see or understand.

In the last decade, a little of this changed. With innovations in genetics, we gained a set of new tools for seeing, a kind of "geneoscope," every bit as powerful and revolutionary as a telescope, though to see the worlds within us rather than around us. This set of tools made it possible to examine the RNA (kin to DNA and intermediary, in your cells, between DNA and protein) found in a sample as a measure of which species are present. One can take a scoop of rainwater and using this approach identify the life in it, or take a sample of feces and at least indirectly look at the slew of genes present and what they tell us about who abounds. Now that microbes can be identified based on their RNA, we do not have to culture them to know whether they are present (though it is still helpful). Such genetic techniques are becoming easy and cheap, so much so that a young student or technician might hope to use them to answer a question of relevance to all of humanity as Amy Croswell, working with her mentor, Nita Salzman, and three other colleagues recently did.

Croswell was a technician in the lab of Salzman, a microbiologist and immunologist in the Department of Pediatrics at the Medical College of Wisconsin. Together, Croswell and Salzman planned the first study of what happens to microbes in our guts when we apply antibiotics. The two and their research group took ordinary lab mice with guts full of wiggling microbes. They then gave some of those mice, and not others, antibiotics. The mice treated with antibiotics received one of a variety of possible cocktails of drugs. The "high antibiotic" mice received a dose of four antibiotics in line with what was "known" by other scientists to kill all the bacteria in the gut. The "low dose" antibiotics mice received a single antibiotic, similar to what your child might receive for an ear infection.[4] The whole project was simple and small relative to the scale of the problem, a mouse-sized gem.

Much of what Salzman and Croswell did was relatively easy. Mice are experimented upon in labs all over the world. Generations of breeding and tinkering have led to cages and protocols that are, if not always elegant, perfectly functional. The mice that Salzman and Croswell would study were the descendants of a mouse family that had been in the lab for tens of generations. It was, in many ways, their native environment, one more like our modern human environment, in some ways, than the environment of their (or our) ancestors. They were born via C-section in the lab, raised on formula, and then, at the age of five weeks, treated with their appointed antibiotics.

Let us pause for a second to consider the possible results of their experiment. Perhaps our collective intuition is that those mice that had been treated with antibiotics would have fewer "bad" bacteria and the same or even more "good" bacteria than they had started with. In the context of human medicine, that is what our hope has long been. What did you think was happening when you took antibiotics? It is always easiest to assume someone else knows the answer, but in this case no one did. At the opposite extreme, other biologists working on mice thought that the antibiotic cocktail that Croswell and Salzman applied should kill all of the microbes. Croswell and Salzman added the antibiotics to the water and waited. After a few days, the scientists then took tiny stool samples from each mouse and, as though to add insult to injury, killed them, sampled them exhaustively, and then whisked them away to a large plastic bin at one end of the lab.

When they looked at the samples from the mice, as expected they found that the individuals that had been given clean water with no antibiotics had a full complement of microbes. Their guts were, like yours, gooey with life, grams of life. The mice that had been treated with antibiotics, however, were another story. The antibiotic-treated mice, mice that by analogy to our own medical system were the medicated, healthy ones, had microbes in their intestines (that is an important result in and of itself), but far fewer, particularly in their large intestines and colons. The effect was greatest for the mice treated with all four antibiotics, but present even in the mice treated with just one antibiotic, streptomycin. In essence, the antibiotics were capable of wiping out billions of cells from those treated animals' guts. While the different antibiotics tended to kill slightly different microbes, none simply killed "the bad bacteria."[5] Many different kinds of bacteria were affected. Since mouse guts are like human guts, this means that when you or I use antibiotics, the same thing is happening to us. When we kill our microbes with antibiotics, we are leaving behind the relatively few weedy species from which a new microscopic empire of life rebuilds. Nobody knew, but now that we do, it seems even more important to understand what it is that those microbes—the ones that we mostly (though not totally) wipe out every time we take antibiotics—really do. The answer involves a young man, a giant steel bubble, and a mistake.

The question of how much and what the microbes in our guts do for us is nearly as old as the study of microbes. Although Pasteur would become a strong advocate for killing the living creatures in our milk (hence pasteurization) and other foods, he believed that the creatures that live in and on our bodies are so necessary that without them, we would die. They evolved, he thought, to depend on us and us on them. Kill the microbes, he said, and you kill the man. In other words, he thought that the microbes in our guts are our obligate mutualist partners, where "obligate" means they are necessary and "mutualist" simply means that they and we both benefit from the relationship. The germ theory of disease, on the other hand, was based on the opposite idea—that some or perhaps even most of the species on us are more likely to do harm than good. No one had ever done the necessary test to see who was right, and yet clearly the answer mattered. In a world in which we continue to scrub off many (though almost never all) of our microbes, the answer matters now more than it ever has. What really happens when you use antibiotic wipes on your hands?

It is worth being reminded here that this question is similar to the one that John Byers asked about the pronghorn: What happens when you take away the predators? It is the same question that Weinstock would come to ask about the worms: What happens when you take them away? It is the same question repeated with different life-forms, by different scientists, as they look at each of the many parts of our bodies.

Born in 1909, James Reyniers, "Art" to his friends, was an ordinary young man—the good Catholic son of a machine-shop worker. He was ordinary until, that is, he became interested, beyond reason, in Pasteur's question. He wanted to know if it was possible to scrub all of the bacteria off a rat or maybe a guinea pig. The idea that every animal in the world was covered in microbes, but that no one seemed to know if they were good, bad, or otherwise irked Reyniers.[6] Rephrased, the question was whether or not the species that live in and on us in great abundance are mutualists (that both benefit from and benefit us), commensals (that benefit from us, but don't otherwise affect us), or pathogens (that benefit from us at our expense). There must be, he came to believe, a yes or no answer, white or black, mutualist or pathogen. No shades of gray were necessary; either the microbes helped or they did not, and if they did not, they could and should be removed. If they did not, then dosing the gut with antibiotics would be good and just. It would be progress, just as the invention of agriculture, the removal of worms, or the taming of the cow seemed to be.

To Reyniers, the problem was a mechanical one. His challenge was to divide humans from germs in the way you might separate gold from sand. He dreamed of germless rats and, with them, grandeur. By 1928 he was convinced he could figure out how to make a germ-free animal. Everyone before Reyniers who had tried to do the same had done so by scrubbing the germs off, one way or another—a kind of Mr. Clean approach.[7] It is the approach each of us uses on our bodies every day and that, in learning that you have trillions of microbial cells on your body (a hundredfold more than your human cells), you might feel compelled to try. Those attempts had failed in the same way they fail you when you scrub. Removing "nearly all" the microbes from an animal is a very different thing from removing them all. From even a single overlooked or persistent cell, billions can rise.

Reyniers, though, was a machinist by training and family tradition, not a biologist, and so he chose a different route. He decided to try to use metal, plastic, rubber, and industrial tools to separate animal from germ. The iron lung had just been invented, as had the first robot. What if, Reyniers thought, he used the same sorts of technologies to construct a microbe-free world, and then allowed mothers to give birth inside that world? Noah put the animals two by two onto the ark. Reyniers thought he could pull them back apart.

If Reyniers could accomplish his goal, he might prove to be the first person in history to produce an animal devoid of germs—bacteria, archaea, protists, fungi, and even viruses. Such an animal would be fascinating and modern. It would also be useful. It would allow scientists to add microbes back, one by one, to understand their effects in ways that had never before been possible. At the time, many hundreds, perhaps thousands, of experiments on guinea pigs, rats, mice, and even chickens had been done in which those animals were given pathogens (such was and is the laboratory industrial complex). But the animals given such pathogens already had on and in their bodies unknowable slurries of other microbes, with unknowable effects. Reyniers thought that he could change what we know about how our own bodies work *and* in the process usher in a new way of studying disease.

It soon became clear that Reyniers planned to do more than just make the first germ-free animal. He wanted to make thousands of them, hundreds of thousands even. Even before he had ever touched a laboratory guinea pig or a rat, he imagined an entire biological realm populated with animals free of germs. It would be a kind of zoo of germless life. When he proposed the project to the faculty and administrators at his home institution of the University of Notre Dame, he argued that it would take fifty years to accomplish, fifty years not just to make the first germ-free animals but to produce them en masse and to study them over generations. Such was his dream, a dream that seems all the more improbable once one learns that he was not yet a full professor. He was not even an assistant professor, a postdoctoral fellow, or a graduate student. He was a nineteen-year-old undergraduate student, a thin boy in man's clothes.

I am not sure what I would say if an undergraduate student asked me for permission to use a big room and thousands of pounds of metal to do a fifty-year-long experiment in which he was going to remove germs from guinea pigs, rats, chickens, and monkeys. None of the responses that first comes to mind includes the words "yes" or "OK." (The phrase "When guinea pigs fly" does, on the other hand, come to mind.) Reyniers, though, was apparently far from ordinary, enough so that when he asked for a room in Notre Dame's Science Hall, metal, and a blowtorch, a dean in the university administration said, "yes." Maybe the dean had not been paying attention; maybe he thought Reyniers was a professor. However those first days played out, they were the beginning. So it was that a boy began what was arguably the most ambitious project in the history of microbiology.

Reyniers's plan was to try to deliver babies by C-section, guinea pigs to start with, without allowing them to come in contact with any germs, including those on his hands, in his mouth, and even in his breath. Reyniers knew that unborn animals, be they human or otherwise, are free of microbes. He thought he might be able to maintain this status, rather than the alternative of trying to kill the microbes once they were established. The animals would then be allowed to live, mate, and die in a world without microbes. Reyniers began what he saw as his life's work aware that if things went according to his most optimistic expectations, he would be at it until the age of sixty-nine.

Using skills he had learned in his father's machine shop, alongside his two brothers, both of whom would become machinists,[8] Reyniers started in on metal structures, great chambers into which he installed gloves for reaching in and doing surgeries. He built many of these chambers, sometimes with his family's help, but often alone. They were half-submarine, half-hospital room. Day or night, when one walked past his small room, he might be seen, welding torch in hand. He was like a sculptor or a visionary artist. He stood back from what he was making every so often to admire his work. "Look at that curve, that smooth airtight seal!" He must have had doubts, but if so they have gone unrecorded. Things failed more often than they worked, often for years. Reyniers was, at least on average, undismayed. Sometimes he even slept beside his creations—a small man next to big metal spheres, each one resembling an Earth.

Some parts of Reyniers's plan were easy to implement. Early on, he found he was able to more or less sterilize the outsides of the mother animals, for example. The mothers would be shaved and plucked (germs love fur, a reality to which we will return), dipped in antiseptic fluid, and then covered in an antibiotic-treated envelope. That much was easy. Anybody could do it, though I suppose it is an uncommon urge. Harder was what Reyniers proposed to do next. He wanted to take the envelope-cloaked mother and transfer her into a metal cylinder where her babies would be delivered by C-section. Making the cylinder in such a way that it would be devoid of bacteria was nearly impossible. The gloves had to be sealed completely airtight, which took fiddling. The gaskets leaked. Then the air inside the chamber needed to be sterilized too. Finally, there was the question of which animal to use. Reyniers tried to use cats, but they scratched through the gloves, breaking the seal of the cylinder and Reyniers's skin, but not his resolve. He was in too deep to go back.

History records only bits and pieces of Reyniers's emotional well-being while all of this was going on. It is easy to imagine that, as he tried to do what he had long planned, he would have become depressed. By the age of twenty, he had not yet produced a single functional chamber. By twenty-six, he had chambers, but had not yet produced a single germ-free animal—though hardly for lack of effort. Many guinea pigs died. Cats died. Mice died. Rats died. Even chickens died. They died because the surgery was difficult and tedious (among other difficulties, in the early years it had to be done through thick rubber gloves), and because at each stage in the process some individuals had to be checked, to make sure they were germ-free. The whole ordeal was exhausting for the animals and for the surgeons/machinists/biologists alike. The odds of failure of any particular attempt were simply far greater than those of success. In these circumstances, I would have gone mad, but Reyniers kept on, and in 1935, at the age of twenty-seven, he succeeded. Based on the successful "production" of a cohort of germ-free guinea pigs, he went public. He did not even bother to write a paper. He just made an announcement to *Time* magazine,[9] and so it was noted that on June 10, 1935, James Arthur Reyniers produced the first germ-free animals in the world. Now the question was whether these germless babies would die.

Reyniers had been working on the project so long that he had, in the process, finished his undergraduate degree and, without a PhD, been hired as a professor.[10] In that time, one would be forgiven for imagining that he had forgotten about the question he wanted to answer when it all began. He had not. The first thing he did when everything was finally working was to compare the guinea pigs in the chambers to those outside. If Pasteur were right, the guinea pigs inside the chambers would die. The microbes in their guts and on their bodies would be so vital a part of them that, in their absence, life would fail.

But the germless guinea pigs did not die, not once he had their diet right. In fact, they seemed hungrier and more active than did those on the outside with microbes. Success! As time passed, the animals in the chambers seemed to live longer too, and they never developed tooth decay.[11] To Reyniers they were a model of what was possible, even for humans. An article in the magazine *Popular Science* in 1960 highlighted the chambers as a futuristic world in miniature, in which animals are no longer susceptible to the whimsy of germs. The matter was, it seemed, settled unequivocally.[12] Mention was made of sending germ-free humans to space, and, if not germ-free humans, germ-free monkeys. The idea that we might make our own living spaces like those of Reyniers's guinea pigs' was so obvious to anyone reading, so implicitly a part of the story, that it barely required mention. Here in these chambers was the future, not just of science, but also of our lives. It was not biodiversity, not the ark of life two by two into the future, but instead the opposite, just us. Reyniers had not only achieved the first set of his goals; he had inspired the imagination of the masses, inspired them to believe that we all might live like his guinea pigs, germ-free and nearly forever.

With time, the scope of Reyniers's work continued to expand, unabated. Notre Dame gave him bigger and bigger spaces in which to work, and then it gave him an institute. He and his father patented a series of germ-free chambers that are still in use, and perhaps most significantly, his approach spread around the world and, with it his animals. Today there are hundreds of thousands if not millions of germ-free animals alive at any time in the world, in thousands of chambers. The chambers have become more sophisticated (they now look more like bubbles and less like submarines), but the basics are the same. They are the descendants of Reyniers's chambers and so are simultaneously nautical and monstrous.

Reyniers was fabulously successful in carrying out what he envisioned at the age of nineteen, thanks both to his own vision and to the capable (and by their own rights visionary) people he hired to work around him, individuals like Philip Trexler, who would go on to make far smaller, cheaper, and easier-to-use chambers than Reyniers's submarines. Reyniers would not live to the age of sixty-nine to see fifty years' worth of his project, but it did not matter. He had succeeded. His germ-free animals would save millions of lives by allowing the study of diseases in isolation from other factors. However, they also had the broader effect of leading biologists all over the world to the conclusion that the microbes in our guts are, on balance, bad. But Reyniers had overlooked something. His oversight was irrelevant (or largely so anyway) for the use of germ-free animals to study diseases. In the end, that was and continues to be their great value. But the mistake mattered when it came to Pasteur's question of what happens when you remove microbes from a guinea pig or, for that matter, a man.

In the context of Pasteur's question—a question germane to any of our bodies of what our microbes do and what we are to do with them—the flaws were not in Reyniers's experiments so much as in his interpretation of his results. Reyniers was a machinist. He was trained in hammers and metal, not flesh and cells. He had no background in evolution, ecology, or any of the fields that would have given his work context. To the extent to which his skills expanded with time, they expanded into management and fund-raising, not the particulars of life. We can forgive him, in this context, for not paying attention to the nuances of his results, ignoring a dead guinea pig or cat here and there. The trouble was that biologists came to see the world of germ-free animals through Reyniers's lens. Reyniers spoke often and with the weight of his institute and accomplishments. His voice came to dominate the field to such an extent that his interpretation became repeated as truth. Each new talk or study added punctuation until one could almost hear it, a drumming chorus of "Kill the germs!" *"Kill the germs!"* and we would be free of our past. Kill the germs and we would be healthier and happier, just like the guinea pigs in their giant metal worlds.

On the basis of Reyniers's work and other work like it, we came to believe that all microbes were bad and so we continued cleaning ourselves, to make our lives more like the lives in those guinea pig chambers. If Reyniers's original experiment was planned for fifty years, the societal experiment, in which we were the guinea pigs, took even less time to get under way. We went from using no antibiotics on our bodies to using thousands of tons of them in just a few decades. The antibiotics were no bubble. They never killed—or kill—all of the microbes, but we imagined they did. The guinea pigs and rats inside the chambers lived longer, and we wanted that too, to be like them. We wanted to step into the chambers of the future, where we were completely removed from the plagues of our past. Such was our confidence in our germ-free future that several children were even raised germ-free (and bereft of physical interactions with other humans) for a while. They were children who lacked immune systems and so, otherwise, stood no chance of survival. We cleaned them of microbes so that they might live at all. We did so on the hope, the assumption even, that such a germ-free life was what we were all headed for. The bubble was, if not necessary, inevitable, a future into which these children would step first. So it seemed.

Reyniers knew of some of the problems with his experiment, vexing realities of life's evolving persistence. It turned out that some viruses are passed from mother to offspring directly and so are impossible to remove consistently. Some life-forms are even embedded in the mother's DNA. In other words, the guinea pigs, mice, and chickens were germ-free, except for those germs of which they were not free.* In the strict sense, there are still no totally germ-free animals, with the possible exception of one strain of rats. More to the point, some elements of the microbial DNA passed from one generation to the next are necessary. Without the microbial DNA in our mitochondria, each and every one of us would die. Our mitochondria are the descendants of ancient bacteria, descendants that live in and help to power each of our cells. At the very least, in this regard, Pasteur was right.

Then there was the issue that even the animals that seemed germ-free did not always stay that way. Every so often a germ of one sort or another would sneak into the chambers. A single bacterial or mold cell was enough to contaminate the chambers. There were and are a thousand ways for such a cell to sneak in, and once in to divide and conquer. Nature loves a vacuum. Microbes love a vacuum-sealed guinea pig chamber. In some cases, perhaps most, the animals did worse when the germs sneaked in. But every so often a germ would arrive and the health of the animals would improve. These differences were interesting, but they also served as

---

*Perhaps, in reading this, you are thinking, "I know how you could get those viruses out." If so, the urge you are feeling is the backbone of scientific innovation, a stew of can do, curiosity, obsession, and a little arrogance. It is of the job of conservative, daily science to suggest that the problems are bigger than you might appreciate. It is your job on behalf of radical, innovative science to go write your own fifty-year plan.

a constant reminder that with improvements in technology there might also come improvements in the ability of microbes, whether good or bad, to get in. At one point, Reyniers lost ten years of his research when a bacterial pathogen made its way into his chambers and killed all of his animals (at which point he remarked to a journalist that he, like most people, did not have many whole decades to lose). It was such sneaky germs that ultimately killed the bubble boy, the most famous of the children brought into germ-free chambers. The bubble boy had been transferred antiseptically into a chamber at birth because he lacked an immune system. Inside his chamber, he was raised by doctors until the age of twelve. At twelve, he wanted out. At twelve, something needed to change and so he was given a bone marrow transplant in an attempt to restore his immune system. The operation went well, and hope held out that this would be a case of the triumph of ego and medicine over disease. But then the boy grew sick. His mother's bone marrow contained a virus, which quickly killed the boy. The persistence of pathogens nearly everywhere, be they viruses, bacteria, or something larger, should have, on its own, ruled out the idea that we might achieve some germ-free utopia for ourselves. We could build bigger and bigger chambers (or larger and larger houses filled with more and more antibiotics), but the larger the world from which we wanted to exclude microbes became, the harder it became to exclude them. What is worse, although the species that sneaked into Reyniers's chambers were sometimes innocuous, the species that sneak past the barriers we attempt to erect with antibiotic wipes, antibiotic sprays, and the like almost never are. Sneaky germs, though, were not the only problem.

One of the first hints at the larger problem came from, of all places, termites. In the dead wood of Earth exists an empire of termites, trillions of individual animals, each of them dedicated to living off what no other animals want. Picture all of the wood and leaves that have ever fallen. Imagine them piling up, rising around you. Most pieces of wood that have ever fallen on the earth have been consumed by termites. By the time of the first mammals, the world was already thick with their nearly transparent bodies and their long, thin, noodlelike guts.

Termites survive by eating what few animals are able to digest, the nutrients in dead wood and leaves, particularly those locked up in two hard-to-break-down compounds, lignin and cellulose. Lignin in particular is a stingy, rotten food. It was long unclear exactly how termites did this good and necessary work. Then, in the early 1900s, Joseph Leidy—forefather of both modern American microbiology and dinosaur paleontology—cracked open the gut of a termite. Who knows what he expected to see, perhaps their food? What he found was a swirling tumult of life, multitudes that looked to Leidy like people pushing shoulder to shoulder out of a crowded meetinghouse. This crowd included bacteria, but also other creatures—protists, fungi, and harder to characterize beasts. These inhabitants of termite guts evolved over a hundred million years traits and behaviors that allow them to be passed from one termite to the next and, in doing so, to get a free ride to wood and leaves. The termites, for their part, evolved guts to help their denizens. In fact, the variation from one termite species to the next tends to be in the shape and chemistry of their guts. This variation favors different microbes in different termite species and, with them, different abilities to digest food. Some termites have microbes that are better able to eat soil, others leaves, still others wood. Some termites, by dint of their microbes, can actually extract nitrogen from the air around them, equally able, in effect, to eat twigs and air.

Just as with guinea pigs, one of the first questions for termites was whether their microbes were necessary. Sure, the microbes seemed to need the termites, or at least most of them did, but did the termites need their microbes? With termites, testing such a question was easier than with guinea pigs. The termites could be warmed or frozen. Freezing killed the microbes, but left the animals themselves otherwise alive. You can put termites in an ice cube tray for a short while and bring them out. They thaw slowly and then look around as if they have been born again. (In fact, in a way they have. They lose all memory of smells when chilled, and so come out unable to recognize their own queen or king.) It is an experiment you can repeat at home, so long as you live where termites live and have an ice cube tray. When this experiment was first performed, there was a surprise, at least in the context of Reyniers's work. When termites are cooled or warmed and their microbes killed, the termites die. They continue to feed for a while, but the food they eat passes through them undigested. They starve even if surrounded by wood. They starve because without their microbes, they are unable to digest their preferred, but difficult, diet.

No one studying germ-free vertebrates (rats, guinea pigs, chickens, and the like) considered this work on termites. In order to do so, they would have had to talk to termite people. Termite people do their own thing. They converse with ant and bee people only begrudgingly, and with people who study people even less. There are a few hundred of them and they are, largely, happy to focus exclusively on termites for the rest of their lives. Nor were vertebrate biologists particularly concerned about termites. Each group went its own way and ignored the fact that the two bodies of research had come to exactly opposing conclusions—one at the cost of tens of years of work and lots of metal, and the other at the cost of an ice cube tray.

The difference in the results between experiments on termite and guinea pig guts is relevant to all of humanity. It explains what Reyniers got wrong in the context of Pasteur's question. It is not that Reyniers made some big mistake, some folly of hubris.[13] He failed in the same way that much of modern medicine does; he failed to put his question in context, whether of our origins or of our modern lives. He wanted to make the germ-free guinea pigs useful by making them survive, which he did. But in the process, he accidentally rigged the competition between germless and germy guinea pigs in such a way that it was nearly impossible for his germ-free guinea pigs to die.

You might pause here to think about the difference between the termite experiment and the guinea pig experiment. The answers are food, disease, and chance. The story of the food is one of plenty, a bounty of last suppers. The termites, when they were cleansed of the species living on and in their bodies, had been given the food they actually encounter in the wild. Without microbes, this food was indigestible. Cellulose requires cellulase and lignin lignase to be broken down into usable nutrients. Termites produce very little or none of these enzymes and so the food they consumed sat in their guts, where any simple sugars were digested, but most of the wood and leaves passed out the other side, smaller but otherwise unchanged. The guinea pigs, on the other hand, had been given as much food as they might ever want. That food was supplemented with every conceivable nutrient until the guinea pigs were nearly assured enough of each and every nutrient. When one diet did not work (which is to say when the guinea pigs died, which happened, it turns out, many, many times), another was tried. The competition between germy and germless guinea pigs had occurred under rules guaranteed to benefit the germless pigs. To Reyniers, the animals were machines to be raced on the best fuel possible. They were like stock cars or steam engines that needed simply to be fueled with whatever they might need. But the guinea pigs, like us, were not machines. The competition that they evolved to compete in, that of natural selection, occurred on a diet of wild foods, not optimal foods, and it occurred in the context of disease.

When Reyniers's experiments are repeated today, in much smaller chambers typically made of plastic, the germ-free animals continue to do relatively well. But there is a caveat. They have to be fed extra food to gain the same amount of weight as the germy guinea pigs, and they have to be fed food that is richer in nutrients than is the food fed to normal animals. The microbes in the guinea pigs' guts, just like those in the termites' guts and, it turns out, in our guts, provide enzymes that their hosts lack, enzymes that allow their hosts to use a larger proportion of the nutrients in their food, particularly those nutrients trapped in the complex carbohydrates one finds in plant material, so-called fiber. The bacterium *Bacteroides thetaiotaomicron*, for example, which is common in human guts, produces more than 400 enzymes related to breaking down plant material, enzymes that you and I lack. When food is limited, the microbes make it less so. The microbes in the guinea pigs' guts and, we now know, in our guts, produce up to 30 percent more calories from food than the hosts can produce on their own. For each food you eat today, this is likely still true. Your microbes help you to get more out of it, more nutrients but also more calories, whether you want them or not.

The second reason that the germless guinea pigs die is that they end up lacking specific vitamins, in particular, vitamin K but also some B vitamins. Without microbes, vertebrates (guinea pigs and humans included) are not able to synthesize enough vitamin B or any vitamin K. Vitamin K works in vertebrate bodies, including yours, to coagulate blood (the "K" is actually for "coagulate," or at least the German version thereof). As adults, we store up the vitamin K that we gain both from eating plants and from our microbes. As newborns, though, we have little vitamin K and, at least at the moment of birth, no microbes. Nor does breast milk fill the void (it is low in vitamin K). Historically, babies relied on quick colonization by microbes for their vitamin K. When newborns do not acquire microbes in their guts quickly enough, they are at risk for developing a disease called, with little in the way of euphemism, hemorrhagic disease of the newborn. Newborns with this disease lack the ability to clot blood and are at a high risk of bleeding to death. As a precaution against this disease, all newborns in the United States and the United Kingdom are now given vitamin K shots at birth. In countries where children do not uniformly receive such shots, hemorrhagic disease is more common in babies that are born by C-section (and hence have less exposure to the mother's microbes during birth) and appears to be increasing in incidence. And just as babies that have not yet been colonized by microbes are at risk of blood coagulation diseases, so too are children or adults who are given antibiotics that deplete their microbes and in turn their ability to produce vitamin K.[14]

If we reorient ourselves a little and think not about guinea pigs and modern infants but instead about early hominids, whether Ardi or her descendants, we can revisit just what it is our microbes once did and whether they were pathogens (as Reyniers thought) or mutualists. They provided vitamin K where it was once scarce, but just as importantly, they allowed us to extract extra calories from our food, up to 30 percent extra. More of those calories would, in turn, have been converted to fat on our bodies, which, for most of our history, was a good thing. In other words, they were our mutualist partners. Most years, but particularly the lean years, their offerings would be the difference between life and death. Most years in our history, we would have survived by dint of our microbes. If one had to spend ten hours a day gathering food without microbes, the gathering day was shortened to seven or even just six hours with microbes. Microbes helped our ancestors to get more from their food, as they had done for their ancestors and so on, going back tens of millions of years. Nor was this even the only big difference between having and not having microbes. There were still the issues of chance and disease, and to understand them, we need to go back to Nita Salzman, Amy Croswell, and their mice.

Amy Croswell and Nita Salzman, you recall, varied the antibiotics that the mice in their lab received. They also, though I did not mention it before, subjected some mice but not others to the pathogen that causes salmonella. Croswell and Salzman wondered whether the native microbes in the mice's guts might help to prevent infection by salmonella, act, in other words, as a kind of living defense system. The native microbes, after all, would have nearly as many reasons to defend the gut as the mice themselves. It was their bread and butter (or, in this case, their mashed-up mouse pellet). The mice treated with the pathogen and antibiotics became sick, but the mice given the pathogen but no antibiotics did not.

When antibiotics were given, the salmonella was more likely to invade their body cavity through the gut. In addition, guts were more likely to be inflamed. But when the native microbes of these mice were allowed to reestablish, the salmonella no longer found its way into the body cavity. It was repelled, apparently by the native microbes that compete with the salmonella and in doing so prevent the salmonella from establishing itself. Antibiotics, in other words, kill the existing microbes in guts (be they ours or those of mice), but make it easier for whoever shows up next to move in. If, by chance, that happens to be a deadly pathogen, the result is dead mice or, in our case, humans.

Perhaps the closest ecological analogy for what Croswell and Salzman observed is the use of pesticides to control fire ants. Fire ants (*Solenopsis invicta*) were introduced accidentally to the United States (and subsequently much of the world) from Argentina early in the twentieth century. When these ants were first noticed in Mobile, Alabama, and observed to be spreading, a decision was made to spray massive quantities of pesticides in the affected regions. In the short term pesticides succeeded in killing the fire ants, but also killed native ants. In the long term, the killing of the native ants appears to have been the more significant consequence. In those areas in which both the fire ants and native ants were killed, the native ants were slow to rebound; not so the fire ants, which appear to have spread faster rather than slower in areas where pesticides had been applied. So it is too that we might expect the invading armies of microbes to advance one antibiotic-treated gut at a time.

Nor is Croswell and Salzman's work the end of the story. In addition to the hundreds or thousands of microbes in our guts, we, of course, have microbes all over our skin, in our hair, and in our mouths. We are covered in many kinds of life. There are fungi even in our lungs. These life-forms are as yet poorly studied, but it remains conceivable that they too help to protect us, or at least that some of them do. That is even more of a problem today inasmuch as we seem committed to using antibiotic wipes on our hands. Recent studies are unable to find any benefit to antibiotics in hand sanitizers, soaps, or other household products in terms of preventing disease. But such products do have disadvantages. They can lead to antibiotic resistance and may also be killing good bacteria and in doing so making room for the bad, which, especially if they are resistant to antibiotics, are all too happy to move in.

What does all this mean for your gut in the modern world? Ironically, and in no small part thanks to Reyniers, we are now more like Reyniers's guinea pigs or a lab rat than was our potential ancestor, Ardi. At least in developed countries, most of us have ample food, and we have attempted, although only partially and fitfully, to make our environment "germ-free." One key difference, though, other than the fact that we, unlike germ-free rodents, remain covered in microbes, is that whereas the food of guinea pigs and rats has been optimized for their health, the same cannot be said of our own diets. In developed countries, the additional calories from our microbes may actually add insult to injury. Worse yet, obese individuals tend to have more efficient, rather than less efficient, microbes, both in humans and in mice, rats, and pigs.[15] In particular, they tend to have microbes that are better at breaking down complex sugars and fats. Scientists have transplanted the microbes from fat mice into skinny mice and in doing so made the skinny mice fat. All of these features of microbes that are efficient at using sugar and fats have become bad now that we have enough food. But, on the other hand, in countries where many (or in some cases even most) people go hungry, which is to say most countries on Earth, the effect of these efficient microbes and microbes in general on harvesting nutrients is likely to still be very beneficial. If you happen to get microbes that are very efficient at harvesting and providing energy from your food and you are hungry, they will save you. If you get those same microbes and feed them chips, cheese, and white bread daily, they are more likely to make you fat. The difference between our modern lives and the lives we once lived has changed the effect our microbes have on us. Once they made us fitter, but now they may make us fatter, though still, it appears, less prone to disease from other microbes.

Living in a bubble without germs is fine, so long as you are alone and someone gives you everything you need, but a leaky bubble and a crappy diet, well, that is another thing entirely. The boy who lived in the bubble, with time, grew terrified of the possibility of a leak in the bubble through which germs might sneak. We have become similarly terrified about germs around us, germs that might leak through the barriers of our antibiotics. But the problem is not the potential of leaks—the problem is our idea that we might create a bubble for ourselves in the first place. Our microbes are largely good for us. Pasteur was right; without their microbes, our ancestors would have died of hunger and disease. Without our microbes today we might be thinner, but we would be missing key nutrients, and we would be at a much higher risk of disease. It seems likely that in the coming years, our frequent use of antibiotics will progressively make each bite of our food less nutritious and give each pathogen to which we are exposed a better chance of taking over our body, one inch of colon, intestine, and stomach at a time. With time, we may learn to better manage for particular kinds of microbes (those that help with vitamin K, but not those that help to make us fat, for example), but that time is not yet here. Nor is this the end of the story. For that it is useful to first go back to other, apparently more sophisticated societies, those of termites and ants and then from them to the human appendix, which (despite its name) is central to who we are.

It would be fair to ask how as scientists and as a society we missed the value of many of our microbes, missed their value and instead of figuring out how to help the good microbes, focused on killing them all. Part of the answer is that there was a time when we were so threatened by microbial diseases that killing them all was not a terrible idea. Reyniers himself may also, in his fervor, be partially to blame. But the big reason, I will argue, has to do with

Babel. A central premise in ecology, and of this book, is that nature repeats itself with variations on a few main themes. If one understands how an individual ecological system, for example deep-sea vents, works, the lessons learned in that system can be applied to others. The crashes and peaks of lynx populations feeding on hares are very similar to those of predatory mites feeding on dust mites in your pillow. By that token, lessons that would apply to our guts can be garnered from the many ecological studies of mutualisms between animals and microbes. Until very recently, researchers studying the human body have missed such lessons, whether they come from termites, ants, tardigrades, at our collective expense, though the problem is not their lack of vision or insight. It may have to do with broader changes that have occurred in science in the last fifty years, changes that find their own best model in Babylon. History, like ecology, repeats. This is why Reyniers missed the real significance of his results for Pasteur's question. It is also why we continue to miss just where it is that we fit in the writhing universe of living things.

In the biblical story of the Tower of Babel, the people of Babylon come together to erect a great building up to the sky. This tower will be their glory, their great and ambitious accomplishment. Brick by brick they raise it, out of sweat-splashed mud. They raise it using their hands, but their common language also helps them to coordinate efforts—to holler "Over here, we need a brick" and to move the tower up, layer by layer. Their language is as necessary to their endeavor as the pheromones of termites and ants are to their success or the dancing of bees to theirs. Language holds them like a thread, but not all that begins ambitiously ends well. God punishes these people for their arrogance by dividing them. He forces them to speak hundreds of languages, an act that pulls them apart. The moral typically taken from this story is one about the consequences of ambition. But there is a second moral too, implicit in the method chosen to divide these peoples—that the failure to communicate leads to failure. Something analogous has happened in science at an increasing rather than decreasing rate.[16] With it, the layers of bricks have grown more difficult to lay. Previous layers exist, of course, on which to stack the mud-baked ideas, but what do they really stand on? More important, where is this tower going? Such answers have grown more difficult to see.

In looking at science from the outside, one might hope that as our total knowledge increases, we gain a broader and more complete understanding of how the world works. Collectively, we may. Libraries grow. But for individuals, it has become more difficult to have a broad perspective. The scientists of each field have developed more and more specific words and concepts for their quarry. It is now difficult for a neurobiologist to understand a nephrologist and vice versa, but it is even difficult for different neurobiologists to understand each other. The average individual's ability to understand other scientific realms has become limited. To do so, they must be scientifically multilingual. Biological polyglots are most rare in the study of humans, where territories are divided finely. One might spend a lifetime studying one kind of human heart cell or some attribute of mucous.[17] The more divided into tiny parts a field is, the less likely some types of big discoveries become. Mechanistic discoveries still happen as scientists struggle, for example, to understand each tiny part of the ear and how such parts come together to make sound, but very few individuals are standing far enough back from what they are looking at to be able to make big conceptual breakthroughs. Instead, such breakthroughs often come first from scientists studying other, more obscure realms of life, realms in which they are still, at least relatively speaking, their own kings and can stand at a distance between bouts of looking deeply. Ecologists and evolutionary biologists are among those more obscure tribes who still step back a little farther (though less far than they once did). From that slightly greater distance, they can sometimes see what was otherwise missed, lost, as it were, in translation. To really see what is going on, one needs to step back far enough to see parallels, the reverberating similarities between one field or organism and another. I would suggest that an ideal distance is far enough away (figuratively) to see both termite bodies and human bodies, but also the big sweep of the ecological world. From such a distance, it is hard to avoid looking at ants.

Ants are, like the other ethers, everywhere. Perhaps *the* classic example of interactions between one species (such as humans) and another species that lives on it (such as our microbes) is that of the relationship between ants and acacia plants. Acacia plants provide ants housing and tiny pearlike fruits in exchange for the ants' protection of their leaves. Plants with ants grow healthier and faster than those without, because in rewarding the ants, the plants gain a defense against another even more costly group of insects, herbivores. In the story of this relationship is an obvious parallel with the story of our bodies and our microbes. But one can also find closer parallels—one need only look to those ants that farm.

Farming ants are more like us than any other species. Farming, or as they are better known, leaf-cutter ant colonies, are colossal societies. They are composed of many thousands, sometimes millions, of sterile individuals, all doing their queen's bidding. Just as in any society, the individuals are imperfect. Some make wrong decisions. Some are eaten. Some carry back toxic leaves. Some persistently walk in the wrong direction. But on average, they get the job done. The job is carrying bits of mandible-cut leaves back to the nest where they are spread as fertilizer on gardens of fungus. The fungus produces sugar-rich bodies—one might call them fruits—which the ants feed to their larvae. For the ants, the fungus serves as an external gut, digesting the leaves in a way the ants on their own cannot. Different leaf-cutter ant species (and there are many) farm different fungi. The ants and the fungi need each other. The tricks the ants have evolved to take advantage of the fungi are many and elaborate. It is not at all easy to farm fungi, and yet the ants seem rarely to fail. Nor is it easy, from the fungus's perspective, to feed ants. Yet the garden grows. The colony grows. The skin around the queen's abdomen stretches ever thinner as she fills with the eggs of prosperity.

Leaf-cutter ant colonies, the culmination of fungus-gardening sophistication, are full of circuslike particulars. Minims, the tiniest leaf-cutters, ride leaves, guarding the ant carrying the leaf they are on from flies bent on laying eggs in their heads. Soldiers, with heads bulbous with muscle at the expense of brain, guard trails. Leaves are cut using a near-perfect, saber-saw-like vibration of the mandibles. And, of course, there is the fat lady, the queen, who, somewhere in a deep chamber, lays thousands of eggs a day, each one as particular and detailed as though it had been produced by Fabergé. Many thousands of tropical biologists have sat with and watched this circuslike civilization proceed down paths. Few have failed to remark at how these cities of ants resemble human cities. It is an almost inevitable comparison, though the ants together also seem to resemble a body. Each individual might be compared to a cell moving around food, shuffling out toxins, doing their unrewarded part to keep the whole alive.

Leaf-cutter ants are remarkable, as are their fungi. Together, they exemplify the extent to which one species can come to depend on another. But biologists studying the human gut did not know about these ants, at least not any more than you might know about them from seeing them on a Discovery Channel special, in and out of focus or in comparison to some human's pinky finger. Also, until recently, the story of the leaf-cutter ants was still missing key elements. It was still unclear how the ants' simple immune systems prevented the fungus, their external gut, from being attacked by disease. (It is a question, you may notice, analogous to that of how our gut prevents itself from being invaded by bacteria that attack our internal guts.) A garden untended tends to be devoured, particularly in the tropics, but these fungi grow relatively pure and, although delicate, untouched. Nor was it clear how the ants themselves, surrounded by fungus, kept from being infected by some pathogen.

In nature, when things go uneaten, there is a reason. They taste awful, have toxins, or are otherwise defended. But what holds back the demons of the ants' gardens and, for that matter, on their bodies, bodies that rub daily among microbes? The answer, it was recently suggested, is "good" bacteria. Cameron Currie, a biologist who is now at the University of Wisconsin, found bacteria living in special divots and rough spots on ants' bodies. The bacteria seem to be more abundant when pathogens are more prevalent in the ants' colonies. Currie has argued that these bacteria are helping the ants to defend themselves from "bad" microbes on their good fungus. It has long been known that bacteria produce antibiotics (most of our antibiotics such as penicillin come, at least originally, from them). The bacteria on leaf-cutter ants may produce antibiotics that repel the bad fungi (called *Escovopsis*) that attack the ants' good fungi. The bacteria, in this theory, are the ants' defenders and partners, farmed by the ants on their bodies, worn like skin on the bone. The ants appear to sustain the bacteria and even to have evolved traits and maybe rewards that keep them from sliding off. An alternative explanation for the bacteria is that they actually defend the ants rather than the fungus. Either idea remains possible. In the meantime, the idea that our bodies might farm good microbes, for our defense, came first from the ants. And because ants are easier to study than humans, the intricacies of the ant relationship (though already contentious) are likely to be resolved more quickly than those of our own. Whether or not he is right about just what is going on, Cameron Currie stood far enough back to notice something interesting, something that applies to ants, but also, as it turns out, to you.

We tend to think of ourselves as complex, or at the very least complicated. In the old telling, we were at the top of the great chain of life. Yet at the same time, we seem to have difficulty imagining that our relationships with other species are as sophisticated as those of, for example, the ants. But our interactions are elaborate too. The leaf-cutter ants were just, until recently, better studied and with more perspective, from a greater distance. We are more like a leaf-cutter ant colony than anyone had imagined, in terms of how we tend our microbial gardens. Our appendices, when they are not bursting, are key to doing just that job. Even as our brains try to tell us that the bacteria in our guts or on our skin are all bad, the appendix mumbles otherwise. In some primitive, wordless way, it knows.

6

I Need My Appendix
(and So Do My Bacteria)

On September 11, 1942, Dean Rector of Chautauqua, Kansas, turned nineteen. The celebration was held more than a hundred feet beneath the surface of the ocean. Above Rector were millions of pounds of water and, more ominously, Japanese destroyers searching for American submarines like the one he was in. The submarine was a chamber, meant to hold back both torpedoes and water. In it, Rector was about to have his first birthday at sea.

Rector's celebration was short-lived. The next morning he woke up thinking he was about to die. Many dangers surrounded Dean Rector and his shipmates, but on this day it was the internal demons that struck first. As the pain grew worse, he began to whimper. It was a sound more like a dog than a man. One sailor thought what was afflicting Rector was "just the flu." "Maybe homesickness," offered another, but as Rector moaned, the reality became inescapable. He had appendicitis.

Under ordinary circumstances, appendicitis can be dangerous, but for Rector the circumstances were far from ordinary. No trained surgeons were on board, and finding a surgeon so far from home and surrounded by the Japanese was out of the question. They would have to operate, but how? Who would perform the surgery? Officially, Wheeler B. Lipes was the ship's surgeon, but in title only. His medical experience consisted of having run an electrocardiogram machine. When Lipes was asked by his commanding officer to do the surgery, he declined. So his commanding officer ordered him to do the surgery. Among Lipes's reasons for hesitation (other than his total lack of experience) were that he did not know how long the ether would last, he did not know where, in a human body, to find the appendix, and he could not imagine, short of kitchen utensils, what to use for surgical tools. Yet Lipes had been ordered and so began his preparations.

After some soul- and equipment-searching, Lipes eventually readied himself to remove Rector's appendix. Rector had been placed on his back on a table in the officer's wardroom. The table "was just long enough so that the [Rector's] head and feet reached the two ends without hanging over." Lipes stood over his patient, still flipping nervously through a medical book (looking, one presumes, for a drawing that would indicate the geography of the offending organ). He was wearing a tea strainer as a surgical mask. The men chosen to assist him had been given kitchen spoons to use as muscle retractors. They stood ready on either side of the patient. Then, as would later be described in a *Chicago Daily News* article, Lipes leaned in toward his patient and mouthed, "Look, Dean, I never did anything like this before." Dean's eyes were wide. He watched as Lipes, following "the ancient hand rule, put his little finger on [the] subsiding umbilicus, his thumb on the point of the hipbone, and, by dropping his index finger straight down, found the point where he intended to cut."

The appendix is the most frequently removed body part. Often, as Dean Rector's situation makes clear, such removals happen under dire circumstances. Go to the office and watch the people moving around you. Few will be missing eyes. None will be missing hearts, but quite a number will lack an appendix. These appendixless many walk by largely unworthy of remark, bearing neither stigma nor obvious consequence. Maybe you are one of them. Whether or not you are, it is fair to wonder: If an appendix causes so much trouble and is, from all appearances, less necessary than a pair of pants (whose absence from a coworker would be noticed), why do we have one in the first place? The answer, it would turn out, has to do with our gut microbes and our evolutionary context. The appendix makes sense only in light of our evolutionary past, though no one aboard the submarine had time to care. They were still looking down at Dean Rector, whose mouth was cracked open in a low moan.

Lipes, having paused to gather his wits, continued to cut.

The appendix is a dangly bit of flesh that hangs from the lower intestine. It is the size of a pinky finger, so while diminutive relative to other organs, it is large enough to deserve explanation. Yet the question of what the appendix does has long had no pithy answer. The heart pumps blood. The kidneys clean the blood and help to maintain blood pressure. The lungs distribute oxygen and remove carbon dioxide. The appendix, well, it hangs. Various talents have been attributed to this small organ over the 300 years since one was first removed from a living human, some magical, but most rather ordinary. It might be part of the immune system. It might have some neurological role. Maybe it is related to hormone secretion or muscle function. The dominant view, though, has long been that it does nothing at all. It is a vestige, like men's nipples or the hind leg bones of whales, a prominent and unnecessary relic of history.[1] That answer is wrong, but until very recently no one knew.

The story of our attempts to understand the appendix begins long before Lipes, but with more speculation than gusto. The primary evidence for the mainstream idea that the appendix is vestigial was that nothing happens when you remove it. That was the sum total of the logic. Surgeons (or, in the case of Lipes, electrocardiogram technicians) have cut out millions of appendices. They have watched the consequences the way you might watch the result of removing an unwanted beam from your house. When the house doesn't collapse, one gains a sense of calmed relief (betrayed only by a mild nervousness when the wind blows). Yet the wind blew and those individuals who had their appendices removed seemed to do no worse and die no younger. Their houses were sound. Had the appendix done something necessary, the appendixless, at least some of them, would have gotten sick.* But just like the guinea pigs in the germ-free chambers, they did not seem to and so it seemed clear that our appendix was a relic of the way our bodies used to work when we were monkeys or perhaps earlier still when we lived, rat-like, among the dinosaurs. Perhaps in our ancestors the appendix did something vital; now it just hangs in our bodies like a clapper in a bell, and every so often, as in the case of Dean Rector, rings out, loud and clear—"I am here. Take me out. *Now!*"

Yet there have always been problems with the hypothesis that the appendix is simply an antiquated, useless, vestigial organ. For one, the appendix kills. Were infected appendices not removed, about half of infected individuals would die, old and young alike. Since one in sixteen or so people suffer acute appendicitis,† were their appendices not removed, then one in thirty-two people would die of appendicitis. If, historically, one in thirty people died of appendicitis, and the presence, size, shape, etc., of the appendix has a genetic basis (which it seems to), then it would not take many generations for genes for big appendices or even the presence of an appendix to disappear.[2] All things being equal, those of our genes and traits that tend to kill us, or even just weaken us, do not do well in the gene pool. Fish that make the evolutionary move to cave life very quickly lose their eyes, because having eyes is both useless and costly.[3] If having an appendix is both useless and costly, then it, like cave fish eyes, should disappear. Cave fish can lose not only their eyes but also the circuitry of their eyes within relatively few generations of going underground. The regions of the brain correspond-

*In fact, the trends are actually the opposite. Individuals without an appendix actually have a reduced risk of some inflammatory diseases. A clue!

†Depending greatly on where one lives. Just as for Crohn's, appendicitis is very rare in developing countries. It may be that it is a modern disease, somehow another consequence of the interaction between our biology and our daily lives. Another clue!

ing to vision dwindle. Not so for the appendix, which—millions of deaths to its credit—remains.

The big problem with the hypothesis, though, is monkeys. If our appendix really were vestigial and useless, we would be able to look at our relatives to infer what its use once was. What did an appendix do in our ancestors when it still had some function, and what does it still do in our close relatives? If our appendix is a vestige of our history, then we might expect monkeys to have more developed and obviously useful appendices than we do. Chimpanzees should have smaller appendices than do most monkeys, since they are our closer kin and their lifestyles (and the usefulness of the appendix to those lifestyles) are more similar to ours. Cave fish may lose their eyes because they are neither useful nor cheap, but one can look to the relatives of cave fish to see what use their eyes had once served. In the same way, we ought to be able to look at our relatives to see what our appendix once did.

Therein is both the heart of the dilemma and the interesting story. Humans and some apes appear to have a more highly developed, larger, and more structurally elaborate appendix than do most other primates such as monkeys, which suggests that the appendix is likely more important to us than it was to our ancestors.[4] The pattern is the opposite of what would be expected if our appendix were vestigial. What could this mean? It seems that the appendix, long seen as useless, does something for us, or at least it did in very recent history. It did so much, in fact, that it allowed those individuals with more distinct and developed appendices to live longer and have more children, who would pass along their genes for a more distinctive appendix. Somehow, we have been getting the story precisely backward. Looking at other primates led to the conclusion that our appendix must, in our recent evolutionary history, have had a value. But what?

For several hundred years, the question of what the appendix did or does hung around, awaiting the right scientist. No one in the world was actively working on the question. Like most questions, it was simply there to be talked about over lunch and then ignored. All across the world, surgeons have spent their entire lives removing appendices, removing so many that the surgery has come to seem mundane, like opening a can of soda or cutting the stem off a tomato. Most stopped wondering at all what the appendix does, not even pausing before they dropped it in the trash. No one had considered the possibility that it played a role in dealing with microbes, but they should have.

Back in the submarine, when Wheeler B. Lipes began to cut toward the intestines of his patient, he knew almost as much as anyone else about the appendix, which is to say very little at all. Knowing he was not the only one who was ignorant would have been little solace to them both. Sweat dripped from Lipes's head and he asked someone to wipe it. A man's body cavity was open below him. The surgery was exhausting. For twenty minutes Lipes searched inside Rector for an appendix, in vain. He "tried one side of the caecum," and then he tried the other. He began to doubt himself.

Then, when all seemed lost, he "got it." Rector's appendix had "curled way into the blind gut." Lipes removed it and put it in a jar, sponged Rector off and, with catgut thread, sewed him back together. Lest it be argued that in any part of the surgery Lipes had actually possessed the appropriate tools, he clipped the thread with fingernail clippers.

Whether Rector lived or died, his appendix was out in a jar for all to consider. Had Lipes taken a good look at the appendix, he would have noticed clues to what it does. He could have seen that the appendix was filled with lymphatic tissue, a sign that it bore some relationship to the immune system. He would have noticed that it was filled with bacteria, a dense carpet of diverse cells that were—like the bacteria on some ants' backs—linked together in a kind of film. He might have also seen that, in the right light, the appendix looked like a kind of cave. Lipes did not notice any of those things, though. At that moment, he could not have cared less what Rector's appendix had been meant to do. He had enough on his mind, which by then was addled with secondhand effects of the ether and adrenaline in which he and the other men had been steeped during the hours of the surgery. The appendix in its jar slid back and forth on its shelf as the submarine rocked in the waves. Rector too rocked in his cot, hoping that he had just been saved.

Over the next days, it became clear that Rector was going to survive. Lipes was his improbable hero. He was to be heralded for the rest of his life for his valor and creativeness. But the more recent story of his patient's or any human's (yours included) appendix features an equally unlikely character, a man who has seen plenty of appendices in jars and trash cans and, in seeing them there, noticed the clues. Randal Bollinger is an emeritus professor at Duke University in Durham, North Carolina. He claims to be retired. Send him an e-mail and his account responds that he won't be coming back to the office until 2050.[*] By standard accounts of the history of science, he is past the prime of his best ideas and innovations. No spring chicken, as they say, nor even a late-summer rooster. But standard accounts have their limits. They tend to ignore the value of experience and observations. Yes, Picasso's best work came out of the wild flame of his youth, but the talent of his friend Matisse was more like a wine, having taken decades to mature. Matisse painted many of his most influential paintings between the ages of seventy-one and eighty-five,[5] and Bollinger, well, he kept working on and thinking about human bodies. They had long been the canvas for his dual arts of mending and discovery. He knew they still had and have a mystery or two left to resolve. One of those mysteries was the appendix.

Through the course of his career, Bollinger had seen thousands of appendices, in bodies, on tables, in jars. He knew the appendix was filled with three things—immune tissue, antibodies, and bac-

*At which date he would be no less than 120 years young. You get the point.

teria. It is the bacteria in the appendix that make its bursting so problematic. When an appendix bursts, the bacteria contained in the intestines and appendix spill into the body cavity and, in doing so, cause infection.

Many people noticed what Bollinger noticed, but most ignored it the way we all must ignore nearly all of what we see. Yet for Bollinger, his simple observations about the natural history of the appendix were about to seem useful. It took the clues he observed and an insight from his colleague Bill Parker at the Duke Medical Center to make the discovery. It was a routine lab meeting in 2005. Parker and Bollinger were talking with students and postdocs about their latest research. The function of the appendix had never been a topic of discussion during meetings, and this day started out as no exception. Parker remembers the lab bench he was sitting at, even which stool. Bollinger, in Parker's telling, "just got this look like he had figured something out," and suddenly said, as if to himself but quietly aloud, "I bet that is what the appendix does." From there, seemingly apropos of nothing, the discussion expanded. The students looked on, excited but dumbstruck. Bollinger and Parker soon believed that they had, in a few minutes on a spring morning, resolved a several-hundred-year-old question. They had figured it out. The answer was suddenly obvious. The appendix, Bollinger and Parker had come to believe, was a house for bacteria. It had evolved to serve as a place where the bacteria could grow, removed from the wash and grind of the intestines themselves. It was a peaceful alley. From that alley, they thought, microbes might also be able to recolonize the gut after an intestinal disease had wiped it clean. Cholera, for example, causes such violent vomiting and diarrhea that much of the bacterial community of the human gut is expelled. For cholera, this effect appears adaptive. When cholera cells are expelled (typically into water supplies), they can then be transmitted to other humans, as predictably as if the water carrying them were a mosquito. Cholera triggers this response by producing a compound in excess that, while not actually toxic, tricks the body into responding as though massive quantities of toxins are present. Under such conditions, perhaps the appendix was a safe house.

At that moment there was little else in the world of which Bollinger and Parker were more sure. Perhaps the bodies of humans were, after all, nearly as sophisticated as those of ants. Now they had to decide what to do. They could try to publish their idea immediately, or they could give themselves some time to better test it. Reluctantly, they decided to wait and test. To do so they would need "to dig up some fresh human colons, with their appendices attached." Although they did not know it at the time, that would take two long years.

The insight that Bollinger, Parker, and the rest of Parker's lab came to that day could not have been made by any of them alone. It depended on their combined knowledge and experiences. It required Bollinger's experiences looking at appendices. Just as importantly, it required a discovery that Parker had made nearly ten years before. Parker was studying antibodies and reading up on how antibodies respond to bacteria. While doing so, he had realized two things: that antibodies sometimes help rather than attack other species, and that the appendix was, for reasons unexplained, full of antibodies. Not only was it strange that this apparently useless organ existed in the first place. It also seemed to be filled with antibodies that the body produces at great cost. Why this might be the case was ignored.

Antibodies are typically described as being part of the body's defensive system, a kind of second line of defense once some other attacker makes it into our body and past, for example, the nose's mucous. This is only part right, though. What antibodies really do is to discriminate our bodies' cells and parts from those of other organisms. From the perspective of antibodies, the world is populated with two kinds of life, "us" and "them," and sorting the two and then triggering the appropriate response is how they spend their lives.[6] The antibody component of our immune system is old. Our immune system works like the system of a rat or a frog, because in the hundreds of millions of years since we last shared a common ancestor, it has worked well enough.

Parker started by reading up on what other biologists already knew about one particular kind of antibody, IgA, which is the most common antibody in the gut. In looking at IgA, he saw what other scientists saw when they turned to the literature. "The primary function of IgA antibodies is to find and identify bacteria in the gut" so that the other players in the immune system can send them packing down through the colon and out of the body. But something about this story did not make sense.

It should be said in advance that much of science is, at least in some detail, wrong. Fixing what is wrong is a big part of what keeps the hundreds of thousands of us who do science busy. The hope is that truth accumulates and mistakes are beaten back (sent down the colon so to speak), but sometimes it takes a while. Sometimes fiction masquerades as fact for generations, growing ever more difficult to see as it is printed in textbook after textbook and memorized by one young scientist after another.[7] Finding these old errors and misconceptions is difficult. But if you can do it, whether through insight, patience, careful reading, good luck, or some combination thereof, it is like nothing so much as finding a door to a secret world right in the middle of Grand Central Station. One wants to stop to ask everyone else, "How did you not see this?"

As Parker looked at the old papers on IgA, he read what every other immunologist had read, but something seemed awkward. The parts were all there, but it was as if they had been put together wrong, the leg glued awkwardly high on the hip. Studies since the 1970s had noted that the bacteria the IgA attacked had a receptor, a kind of microscopic door for the IgA. When the IgA attacked those bacteria, they did so through that door. What, though, were the bacteria doing with a door for the very antibodies whose goal it was to attack and expel them from the body? It was as though the Chinese, after having built the Great Wall, had also left out a giant ladder. Why offer your enemy a door? Then, as Parker read on, he found something even stranger. A recent study had shown that in patients and mice that lack IgA, bacteria with receptors for IgA seem to disappear.

Parker is a medical researcher, studying the xenotransplantation of organs from one animal to another. His job is to find medical solutions, breakthroughs, and applications. Understanding IgA and related antibodies had started out for Parker as a means to an end. If he could temporarily block or change their action, he might get the human body to accept a monkey's lung or a pig's heart. (He envisioned, at some moments, the headline "Pig Lung Transplanted into Cleveland Man. . . .") But in addition to being a medical researcher, Parker has an eye for radical new ideas. He loves them. They rise in him like joy. Now, just as he was supposed to be really buckling down to understand xenotransplantation, a radical new idea was rising in him about IgA. If he was right, whole chapters in textbooks were going to need to change.

So it was that in 1996 Bill Parker found himself sitting in his lab, thinking about what he knew about IgA. Many scientists have these moments when they mentally paw over the facts—like a jaguar with an armadillo—to try to sort out explanations of how things fit together. More often than not, the armadillo proves too difficult to open. It continues to walk around the lab, taunting the researchers, but every so often the jaguar finds a soft spot. Parker thought he had found a way in. He had an explanation that made all his disparate observations make sense. The answer had been there the entire time. It had not even required any new observation, at least not yet. It was a theory that, if right, would change everything we know about the most common antibody in our gut.

In 1996, the epiphany that Bill Parker had was that if his or any other body was, through the production of IgA, trying to get rid of or otherwise control these bacteria, it was doing a really bad job. If the bacteria were trying to avoid the IgA antibodies, they were similarly ineffective. The bacteria had not only left the door open but changed the lock so that it would better fit IgA's key. In fact, it wasn't just that the bacteria had a door, a receptor, for the IgA; the reverse was also true. IgA antibodies have sugars that bacteria recognize and respond to. What Parker thought was that everyone who had ever studied IgA in the gut had been wrong about its function. The IgA were actually helping bacteria! They were helping them, more specifically, to clump together and to set up shop in the gut without getting washed away.

The IgA antibodies, he imagined, help the bacteria by providing a kind of scaffolding with which they can link together to form biofilms, a sort of commune of unrelated microbial cells. Biofilms are common in nature. The bacteria that live on leaf-cutter ants form a biofilm. Bill Parker did not yet know about the leaf-cutter ants, but he did know about similar interactions in plants. Parker's realization was that the bacteria in human guts were remarkably similar to those on plant roots. What if the human body, like plant roots, was producing compounds to help the bacteria adhere? What if IgA was, instead of fighting the bacteria, helping some of them stay put?

In order to test his idea, Parker needed to establish a system for studying IgA interactions with gut microbes in the lab. He began to grow gut cells on a kind of filmlike plastic and layered microbes on those cells. All around the lab were flasks of the awful-looking scum, in some cases primed with human feces. Sometimes a big discovery involves pushing a trail through to a new forest, where great marvels suddenly become apparent. Other times, though, it is a lab filled with bacteria from human poop. To all the world who cared to look, this scene seemed terrible, a little vulgar even—all the world, that is, except Parker, to whom the lab smelled like discovery, potent and yet somehow sweet.

In 1996 all Bill Parker had was an idea, but it seemed to be one with wheels. He needed to test this idea. Those tests would come, albeit slowly, over seven years. Eventually, Parker was able to show in the lab that when IgA was added to biofilms, they formed faster and grew thicker. Bacteria were nearly twice as likely to stick to human cells when IgA was present. When an enzyme was used to break down the IgA, the biofilms fell apart. Yet even when he thought he had the data to support his idea, for a while no one believed him. His wheels spun. No one funded his grants or printed his research papers on the topic. Eventually, in 2003, he was able to publish his paper. But would anyone notice? Or would it drift among the annals of obscure ideas about life? Parker could be both right and ignored.

Then, finally, in 2004, a breakthrough came. Jeffrey Gordon, a more senior scientist than Bill Parker, with millions of dollars in research support and nearly a dozen postdocs, wrote a conceptual paper that supported Parker's idea.[8] Gordon's paper seems to have been the threshold, the necessary traction, and soon others who had been quiet had been given permission to believe. Almost as quickly as a tide climbs up under the roots of mangroves, Parker's idea went from heresy to, if not dogma, credibility. IgA, it now seemed evident, performs the primary function of helping bacteria. In Parker's lab study, the one that initially no one would publish, he found that gut bacteria grow fifteen times as fast when IgA is present than when it is absent. So not only did IgA allow bacteria to do better—it allowed it to do much better.

The transition Parker's ideas were quietly ushering in was revolutionary. When Parker began his work, it was believed that the function of the native immune system in our gut was primarily to attack bacteria. Case closed. Now Parker and a growing number of other scientists argued the exact opposite. Our IgA antibodies, in knocking on bacteria's doors, do not attack them. They help them by providing substrates necessary for them to link to each other and form the commune of bacteria called a biofilm. Such biofilms, Parker and Bollinger have gone on to show, appear to line much of the lower gut, particularly the colon and the appendix. They look, in cross section, like a rug of rod-shaped forms, side by side, tiny soldiers, shoulder to shoulder.[9] These biofilms are often thought of as "bad" in a medical context. They grow along the insides of tubes and on equipment. But in our guts, they might not be bad. In our guts they may be good, even necessary. We will return to the question of "good for what," but first, Bill Parker was not quite done thinking.

It was with Parker's idea (by now, fully a discovery) in mind that Randal Bollinger offered his hypothesis about what the appendix does. If the immune system was helping bacteria in the gut, and if the appendix was where immune tissue and antibodies were most concentrated (and where cells were sloughing less rapidly, making it a kind of slow pool, rather than a river), it seemed the appendix might be helping bacteria disproportionately. The appendix is a small incubator, removed as it is from the fast flow of the intestines (and the potential from infection by passing pathogens), a Zen garden of microbial life.

Bollinger and Parker needed to wait for colons in order to confirm Bollinger's instinct that biofilms were dense in the appendix. When the colons arrived, this is in fact what they showed, a kind of miniature forest of cells—a nest, you might even call it—of life. Bollinger's interpretation is that the appendix houses large numbers of bacteria in biofilms that in turn offer services to our gut, for our benefit. In addition, Bollinger argues that the appendix is a shelter from the storm. When a severe pathogen such as cholera wipes out the good bacteria of the intestine, they can, from the appendix, be restored.

For now, what Bollinger, Parker, and colleagues have offered is the only hypothesis that makes sense given what we know about the appendix. The hypothesis explains other patterns that were associated with the appendix but poorly understood. It explains why appendicitis is common in developed but not developing countries, where humans more often get very sick, particularly from intestinal parasites. Such a pattern is what would be expected were the appendix, in developing countries, still performing its role of replenishing the gut. In contrast, the appendix in developed countries is infrequently challenged by pathogens. It may be understimulated (much as the immune system is more generally in the absence of parasites and/or pathogens) such that appendicitis is, in its way, like many other modern diseases, the consequence of the loss of species from our daily lives. When our appendices burst, it is because our bodies are turning on themselves. As for Dean Rector, the first patient of an underwater appendectomy, although he lived, tragedy would still find him. Later in the same year, Dean Rector died when a torpedo that his submarine had fired malfunctioned and did a U-turn. Like his appendix, it came back at him. This time, he had no chance to get out of the way.

It was a breakthrough to realize that our immune system, appendix included, might be helping rather than hindering the microbes in our guts. It reversed the false conclusions of decades of research in giant guinea pig chambers and suggested not only that we might benefit from our microbes but that over evolutionary time we had benefited so much that it was worth evolving specific antibodies and organs to make sure those microbes were treated well. Alongside this change of perspective ushered in by the work of Amy Croswell, Bill Parker, Randal Bollinger, and others, a whole new field of inquiry has emerged. For most of the history of human medicine, we have thought of other species as negative.[10] Bacteria kill us. Fungi kill us. Worms, viruses, protists, and the other legions of biological doom kill us. What had prevented medical biologists before Parker (and others who were at the same time coming to similar opinions) from seeing what he saw was, in part, this vision of other species as deadly. At best, other species were seen as having no effect on us. But they could not be helping us. Mutualisms were reserved for the ecologists studying obscure organisms (tropical ants and termites, for example) in faraway lands.

The story of the gut, the appendix, and their bacteria is the tip of the iceberg, and we are only just beginning to see what remains hidden beneath the surface. Our bodies have adapted to interact with many species other than bacteria. You and I are like the colonies of leaf-cutter ants, dependent on other species without which we would not be entirely whole. We imagine ourselves besieged by germs, but this is a mistake. Our bodies are integrated with the microbes. In cross sections of our guts, it is impossible to say where the bacteria end and our guts begin. The IgA antibodies fail to recognize our good bacteria as foreign. Those good bacteria are, to IgA, self, the same as any of our other cells. While this new view of our lives is foreign to the medical community, to ecologists it is familiar. It is more the average way of living than the aberrant one.

It is hard to visualize the interactions of our bodies with other species, and for the immediate future it seems likely to remain so. We can imagine our guts and even the diminutive house that many of us are, right now, providing to bacteria in the form of an appendix, but even this is vague. Re-creations of the nests of leaf-cutter ants, on the other hand, offer a side window into what we might see. Recently, scientists unearthed an entire colony of a species of leaf-cutter ant. They poured gallon after gallon and then truck after truck of cement into the nest. The cement, mixed wetter than normal, poured down every tunnel into the great city of ants, killing first the workers, then the brood, and finally the queen, but in the process freezing for perpetuity a negative version of the construction. Other ant nests have been cast in this and similar ways, but none this big, none even close.

After several days of pouring, the cement finally filled the nest. It was allowed to harden and then, resembling nothing more than ants, the men began to excavate the soil around the now solid cast. As the cast was slowly revealed, its tunnels and chambers began to emerge. The scene looked like an archaeological dig. It looked like the Chinese excavations of the terra-cotta warriors, their heads, and then their shoulders and then their entire bodies emerging from beneath the sand. The workers kept digging. More tunnels. More chambers. And on the fifth day, the nest was revealed. The hole was ten feet deep and twenty feet wide. The nest's form was that of a heart. The beating center, once filled with ants, was now stilled, and around the center were the arteries and veins and chambers of life. With patience one could see details: the garbage rooms, the dangling vaults of fungus, the queen's deep quarters. It was all there. The cast had worked nearly perfectly. It was a work of art, the product of both ant and man, though mostly ant. Other smaller casts of ant nests have hung in museums. This one was too big, far too big. The men sat around it, though, as if in a gallery, stepping back to gain perspective, and then coming closer for detail. In that moment, they had something ever harder to get in science, perspective.

The physical nest of the ants has been sculpted by evolution to benefit not only the ants but also their partners. Special tunnels have arisen to ventilate the fungus. Chambers have come to be shaped so as to facilitate the fungi's growth. Garbage chambers or piles are placed distant from the fungus, so that any pathogens that spring up on the dead might be isolated and controlled. Our body is not unlike an ant nest, composed of multiple cells and multiple species. What is surprising, though, is that while we are surprised but not shocked by the complexity of the relationships between ants and microbes, we don't expect the same of our own bodies. We have no trouble believing that an ant colony depends on the complex microbial slew growing on individual ant bodies, in their guts, or covering and including their fungus. We have no trouble believing that subtle changes in the plant communities around ant nests are enough to fundamentally alter a colony. That the same is true of our own lives is somehow harder to believe. We think of ourselves as complex animals, sophisticated even, but somehow we imagine that a complexity of interactions is reserved for other species.

The appendix is a window into our similarity to the ants and other life-forms. Open the appendix and examine its contents. Spread them around. They are messy but they can be read. Their message seems to be that we have evolved, unique from our closest relatives, a special appendage to house bacteria, filled with IgA antibodies that help them to hold on. The appendix and IgA antibodies are a metaphor for our bodies more generally, bodies that fight some species, but also, whether we are conscious of it or not, have evolved special abilities to help others, some of them as small as a bacterium, others, it turns out, as large as a cow.

PART IV

How We Tried to Tame Cows (and Crops) but Instead They Tamed Us, and Why It Made Some of Us Fat

7

When Cows and Grass Domesticated Humans

Out with the bad, in with the good. We tend to think that the changes we have made to nature have disfavored species that harm us and favored those that benefit us. One might hope so anyway, but it is not the case, at least not universally, maybe not even on average. The species our brains urge us to disfavor are sometimes species that kill us, as with the nastiest of our worms and many of our microbes. But the species we disfavor also include most of the fruits and nuts we once harvested. These were the sweet and fatty species that sustained our evolution, species that would have touched Ardi's rough lips, species once treasured, but now ignored.

For much of our primate history, we spent hours a week picking and savoring wild fruits. The fruits benefited us. We also benefited their seeds by "depositing them" wherever we relieved ourselves. Some plant species spread around the world in this way, using latrines as stepping-stones. In this regard our ancestors were like toucans, emus, monkeys, and the many other species that serve plants as seed dispersers. We ate other things, of course. We searched out some insects—the queens of ants, for example, or the grubs of large beetles—but, for most of our story, the plants were the mainstay of our vessel. Today when we look out at our evolutionary partners, the ones not in our bodies, we see a very different scene. No less than half of all wild forests and grasslands have been replaced by agricultural and other intensively managed land uses. On these managed lands, we nurture a tiny minority of Earth's species, our domesticates, whether corn, rice, wheat, or, more rarely, something else. These species are still our mutualists, but in a very different way from the papaya tree growing like a phoenix out of the outhouse. In making the transition from gathering thousands of species to farming far fewer, we caused both our favored species (domesticates) and our disfavored species to evolve, but they were not the only ones. We evolved too. The story of just how we changed begins with the earliest days of farming and what, from that seed, would ensue.

From a distance, farmed land seems possessed of both power and beauty. In early landscape paintings, fields glowed, pregnant with seed. But farming is a dark art. Bad years and hard days outnumber the good or easy, and yet we have to persist, regardless, because there is no longer an alternative. Once, we could find all the food we needed by simply walking around, searching. A hundred thousand years ago, all humans lived in Africa. Then, one lineage of humans (one branch on the human tree) left East Africa and made it to Europe and then later down to tropical Asia, Australia, and eventually North America. In all of that time, no one farmed. Everyone learned the species around them and collected and killed them. Then, over the last 10,000 years things began to change. Agriculture arose, multiple times in separate regions, and spread as it has continued to do ever since. Nearly all of the food consumed by humans is now farmed, whether in a field, pasture, or cage.

It is easy to forget how recently the world was otherwise. In the Amazon, for example, just 6,000 years ago, groups of humans were once small in number and lived along beaches, beside rivers, and under the canopies of trees. They gathered what they needed. Such groups would have stretched from Bolivia to Ecuador, spaced out and separate. These early settlements in the Amazon have been poorly studied. Bones and fossils are quickly broken down by roots and reinvigorated into leaves, termites, and beetles. Yet because the Amazon was colonized more recently than most of the tropical world, it provides as clear a portrait as we have of the transition from who we were to who we are. What we know is that once the Amazon was colonized, groups moved along rivers. They moved to the best spots and then, when the best spots were taken, beyond them. Year by year, more groups emerged, each with more individuals. The Amazon is immense, but finite. Wars kept some check on populations, as did lean years and infanticide. But eventually the Amazon abounded with humans running along under the trees. In each village, both in the Amazon and everywhere else on Earth, people would have learned the plants and animals around them, not all of them, but many of them. It was, in some real ways, a golden age of knowledge. Contemporary indigenous peoples living in tropical forests tend to know hundreds of plant species and perhaps a similar number of animal species, a large proportion of which are used, whether for foods, medicines, construction, or even as toys.[1] If the same was true of their (and our) ancestors, hundreds of thousands of species might have been known and used globally. Collectively, our gathering forebearers knew the uses of more species then than we do now. They did not know about the germ theory of disease or particle physics, but they could distinguish the tasty fruits from the deadly ones and knew enough about the biology of each edible animal to know how, when, and where to chase it down.

Yet even though the indigenous peoples of the Amazon and elsewhere could extract nutrition from many different species, the growth of the forest and its creatures was not limitless. At least in some tellings, the Amazon, like the Congo or the forests of Asia, was a kind of petri dish, bounded by the Andes on one side and by oceans and deserts on the others. In this flat dish, populations grew more and more dense until there were millions of people in the forests, all of them gathering fruits, and killing monkeys and birds.[2] Think for yourself about the possibilities in such a scenario. Populations would have grown until resources were depleted. And then what?

When human populations grew in the Amazon and elsewhere,

they might have simply met with increasing rates of death and war. This is what happens to bacteria. It is the reason that we are not feet deep in great piles of microbes. Or they may have spilled over into marginal lands, farther from necessary water or easily accessed food. These were possibilities undoubtedly encountered in some places. The other possibility, though, was that some populations might find other ways of surviving. In such circumstances, two "inventions" repeatedly appear in history: agriculture and civilization— bread and kings.

As you think about your own life, it is worth wondering what effect the transition to agriculture had on you. It is worth wondering what kind of success came of being the ones who survived, no longer by the many fruits, nuts, and animals but instead by the few that would grow despite being harvested, the ones that might be tamed and that you now either grow in your garden or buy processed at the store. What happened, in other words, when history took the wild species of our original diets away? The answer depends as much on who your ancestors were and how their diets changed as on the simple fact that diets did change, more slowly in some places than others, but nearly everywhere eventually.

Let's return to the Amazon. Across much of what we now tend to think of as the "pristine" Amazon, civilizations and agriculture once flourished.[3] They flourished at the margins of the forest in seasonal lands where good years are good, but bad years are very, very bad. In those places, populations grew denser than elsewhere. Villages became great cities of thousands and then perhaps hundreds of thousands of individuals. One can fly over the Bolivian Amazon today and see the ruins of these civilizations, hundreds of miles of earthen, raised roads, the grid-iron prints of raised fields and everywhere among them, house mounds—the crumbling ant hills of man. Separate and similar civilizations arose in Colombia, Peru, and Brazil in concert with the development of agriculture. Peanuts, manioc, and sweet potato were farmed on long raised fields, separated by floodwaters. Other crops would grow at high elevations, where the Incas seeded their empire. But all across the region they grew, and as they did, people stopped moving their houses. Lifestyles changed. Humans went from who we were to something more like who we are, settled, agricultural, and dense.

Similar transitions would happen elsewhere in the world. We invented agriculture multiple times much the way a single storm might light many fires. It is typical to see this transition from hunter-gatherer to agriculturalist as one of our greatest human successes, the raging light of bounty. With agriculture would come complex societies and their trappings, writing, art, music, and the sort of unimaginable intricacy of culture we now confront in our daily lives. In many cultures, agriculture is afforded the special status of a blessing or rebirth. In some Amazonian groups, the first people are said to have come from a cassava plant whose roots, buried just right, sprouted legs, arms, and soul. Demeter, the Greek goddess of crops, brought springtime and youth. She and agriculture were, in no uncertain terms, the signs of our reemergence as a species that could change the land in such a way as to make it more bountiful. Though many insects have evolved the ability to farm, among the mammals we are unique. We learned to farm after the farming ants, beetles, and termites, and so now, like them, we sow the fruits whose consequences we reap.

It would be easy to imagine agriculture to be at the root of our health and happiness. It is not. For one, with the transition to an agricultural lifestyle (from a dependence on many species to dependence on a few), life expectancies tended to decrease rather than increase. Hunter-gatherers seem to have lived longer, on average, than did early agriculturalists. In addition, various measures of "wellness," the kinds that can be discerned from bones anyway, also worsened. The transition from hunter-gatherer to farmer tended to lead to an increase in disease burden, including digestive disorders associated with this new diet. Worse yet, the new agricultural diets were coupled with social hierarchies and haves and have-nots, so that even when there was enough food, not everyone received it.

More than ever, with agriculture, survival came to depend more on status, culture, and the complexities that emerge from thousands or even millions of people living together than on avoiding predators or simply foraging for enough food.

One begins to wonder why, if agriculture had so many negative effects, we chose to engage in it in the first place. Farming is hard and unhealthy. So why do it? The answer, at least in part, may be that in many of those key moments in history when our ancestors chose to farm, the choice was not between the good days of hunter-gatherer society and agriculture, but instead between the worst days of hunting and gathering (when the food had run out) and agriculture. At least this is what Leigh Binford, an anthropologist at the University of Connecticut, proposed in the 1970s. Binford imagined that each of the many upwellings of agriculture came about when communities found themselves without alternatives—out of food and options. He said this not as a crackpot outsider, but instead as one of the most well-regarded thinkers in anthropology, which does not mean that he was always right or that other anthropologists have always agreed with him. Yet if he was right, it might mean important things for who we are. If each agricultural people started not from a great empire but instead from a few who struggled and made it, then each group of agriculturalists might have unique genes, genes of happenstance or even, just maybe, genes that allowed them to survive on their particular new crop or animal. Many of us might then descend from these small groups of individuals, the lucky few with their particular crops and genetic variants, the mutants who made it through.

Binford and his critics agreed upon some parts of the story of the origins of agriculture. It was known that long before agriculture was practiced for sustenance, it was done more casually. Someone might find a vine he or she liked and sink it into the ground by their house. They might uproot a favorite tree and replant it. They might even more actively cultivate a few things here and there.

That was the extent of it perhaps for the simple reason that harvesting food from the forest was easier. It did not require much time, just four to six hours a day, to collect enough food for a family.[4] The rest of their time was leisure. On this point, there is little disagreement. The average day of a hunter-gatherer was composed of a little work and lots of time for art, dance, and, one presumes, sitting around and telling stories. The disagreement begins in terms of what happened next.

Imagine for a moment the scenario as Binford saw it. You live in a small community of a larger tribe that over generations becomes larger and larger. As it does, food starts to become scarce around the village, particularly as movement becomes more difficult. Favorite plants disappear first. Pests build up. Fleas live in each house, lice too, and other animals. In times gone by you might have moved your village. Perhaps you would have moved it sooner in some places, such as the tropics, than others, but eventually, regardless of where you were, you moved it. Innumerable villages in the Amazon, as elsewhere, roamed like locusts. In the Amazon, it takes about fifteen years for the fleas, lice, bats, and other realities of jungle life to build up in a group of houses beyond livable densities. So it was that most Amazonian hunter-gatherers, just like those in tropical Africa or Asia, once moved about every fifteen years. But at some point you could not move anymore. The forest in every direction was already occupied. So you stayed, and as you did, the inevitable sank in around you. Pathogens and their diseases grew more common, and the food on the trees and in nests and caves grew less. It happened, perhaps, in a bad year, when there was not enough to eat in the other villages either. In such years, many would have died. They would have rocked in their hammocks, hungry and flea-bitten. Those who survived may have been the few who still had some food, food planted near their houses, not yet domestic crops, but crops that might be domesticated. They ate what grew and, if they could, planted more. Whole villages died, but in your village the few plants took. On those first crops—the near-cold embers of humanity—you and a few others survived, tending each stem and seed carefully, the way one might tend a fire in the dark, with the knowledge that each seed must be passed on, back into the soil, as they have been until today.

Binford imagined that at least some of our ancestors in the Americas and elsewhere turned to agriculture out of such desperation. He imagined that as populations grew too dense, many individuals died from starvation. Perhaps only a few individuals or families would have farmed enough, quickly enough, to survive. In Binford's view, agriculture was, at least sometimes, and maybe even usually, a desperate act. From that act arose societies that had not invented agriculture, but rather had been conquered by it. The first crops would have begun, almost by definition, as too rare to sustain anyone, or nearly so. They would have been difficult and tedious. Any variants of genes in humans that made it easier to live off those crops, to digest them, for example, would have been favored. So it was, Binford imagined, that their cultures and maybe, just maybe, their genes, came to revolve around agriculture. Communities became sedentary and their diets grew worse. In agriculture, Binford saw a bane of humanity. But by the time humans began to write, we had committed ourselves irrevocably to that lifestyle. If correct, this view would have many consequences for who we are and how we live now. Among the consequences is that crops are inevitable, permanent, and problematic. Crops allowed a few lucky peoples to increase in density beyond what had ever been possible before, but whereas once they had depended on thousands of species as our mutualists, they came to depend on just a few, and in some places, just one. They and, by descent, we became linked not just to food production in general, but even to individual kinds of crops. In the Amazon, it was peanuts and yucca (*Cassava manihot*). Elsewhere this tragic play would be reenacted with different species, but with actors and roles that were, to Binford, eerily the same.

Binford's view of agriculture as a kind of postapocalyptic sustainment was attacked and criticized by other anthropologists for years. It conflicted with the story we tell about ourselves in which we are innovative, successful, and in control of our fate. With time, Binford went on to do other things and think about other questions. He had, he thought, good evidence for his theory, but some potsherds and bones were not enough, and so his idea lay fallow, until now.

Binford's view of history and agriculture may not be the rule everywhere. Recently, the evidence seems to be gathering that it is the rule, or at least the story, some places. A telltale gene, a gene that points to what happened to humans at the dawn of agriculture and how we changed, has now been discovered. These genes and their variants prove how important the domestication of plants and animals was to our survival, if not in every case, then at least sometimes. They also make clear what Binford could only allude to, the extent of the domestication of humans.

The story of these genes begins with the aurochs (*Bos primigenius*), the ancestor of modern cows. Aurochsen evolved in northern Africa and southern Asia at a time when grasslands had spread across much of those continents and tropical forests had shrunk back toward the equator. These beasts fed on the new, more bountiful kinds of grasses. They were just one of many kinds of cowlike animals that evolved during this time, including bison (*Bison bison*), banteng (*Bos javanicus*), and others. Yet they would be the chosen one, one of our chosen ones. Each adult aurochs (plural aurochsen) stood more than six feet tall at the shoulder. Aurochsen had the shape of a cow, but were closer in size to a small elephant. Oversized, they spread across the expanding grasses like aphids over the canopies of trees. Their big teeth evolved to clip the grass toward its base and grind it. Inside their stomachs was a storm of life—bacteria, archaea, even protists—which aided them in digesting their food. Without these, their own partners, they would not have survived.

Aurochsen prospered on the great miles of grass. Yet, as much as the green buds were sustaining to the aurochsen, such succor was elusive for humans. Humans never evolved the ability to eat grass. We can chew it, but the cellulose and lignin out of which it is made passes through us, untouched by our digestion. Only the seeds, the so-called grains, satisfy us, and so, at least initially, in looking out at the fields of green, our ancestors were parched mariners at sea, damp and still thirsty, surrounded by food and still hungry. Yet each food must be tried before it can be abandoned, so our ancestors would have grabbed handfuls of grass and stuffed it into their mouths. They would have chewed and hoped. We failed, precisely where the aurochsen seemed to succeed.

Success though, just like failure, has its limits, its breaking points. Eventually, the aurochsen ran out of new grasslands to colonize. Grass stopped at the forest edge, and so did the aurochsen. Limited by such natural boundaries, they seemed as successful as they ever would be. Then, on some dark path, an aurochs met a human. A deal was about to be made. In the real story of temptation and consequence, the forbidden and consequential fruit was not an apple, but instead an aurochs's hairy teat.

In the beginning, the relationship was awkward. There was fumbling and a kind of inescapable nakedness. Initially, milking would have been difficult. Even today, considerable effort goes into convincing cows to "let down their milk." As Juliet Clutton-Brock puts it in her book on the natural history of domesticated animals, "The cow must be quiet, relaxed and totally familiar with the milker . . . her calf must be present, or a substitute that she identifies with the calf . . . and it is often necessary to stimulate the genital area before the milk-ejection reflex will initiate secretion."[5] Awkward indeed. Yet from that first human beneath the cow and others like him would descend much (though importantly not all) of future humanity.

Just as with the origins of agriculture more generally, the details of this beginning were long speculated upon, but difficult to know. Many archaeologists and cultural anthropologists imagined that such domestication events, be they of cows or crops, happened at the margins of an already successful society. They believed the story as it had been laid out in textbooks. The domestication of the aurochs was yet another manifestation of our power for innovation and the control of nature, a kind of nascent science and technology. In retrospect, our ability to shrink wild animals and make them live with us seems near magical. We tamed the cow, but also, later, horses, goats, cats, and dogs too. We did it one coupling or slaughter at a time. It is easy to get caught up in this godlike act of transformation. It turns out, though, that the aurochsen were not the only ones that were transformed. They changed us too, though no one knew that until very recently. They did not "intend to" and yet the result, from evolution's unconscious perspective, is the same as if they did, as if they wanted to be ours and us theirs. Even Leigh Binford, who thought that agriculture was the result of survival through hard times, did not imagine how hard the times might be, nor how much they would change who we are.

In the last five years, the modern tools of genetics have allowed us to ask not just how species are related, but also when particular genetic variants and the abilities associated with them evolved and, once they evolved, how quickly they became common. Most new genes—mutant genes—disappear immediately. Their bearers die. A few new genes persist, and those that persist meet one of several fates. They may persist for a while and then disappear. They may increase, slowly, in their commonness because of some modest advantage they confer. What is also possible, but unlikely, is that once a new gene arises, it becomes nearly immediately universal. The only way for that to happen is if its bearer mates with nearly everyone, or if nearly everyone missing the gene dies. Geneticists euphemistically call such scenarios "selective sweeps," which sounds more like a good hockey series than a period during which individuals without a certain trait fail.

New genetic methods have allowed scientists to reconstruct the stories of the aurochs and of humans separately and then to weave those two narratives together. The story of the origin of the cow from the aurochs begins around 9,000 years ago, somewhere in the Near East, where aurochsen began to come into human settlements. Perhaps that is where they found the sweetest grass. Perhaps that is where they were safest from predators. It is difficult, maybe impossible, to know. Yet, however they initially came to us, whether easily or stomping and kicking, in relatively few generations they were tamed. From this and subsequent points of contact, the aurochsen would spread with humans into new areas, including much of Europe but also new habitats in Asia. They spread into habitats as humans cleared them. They spread beyond their native range, and as they did they became more and more different from those individuals untended by humans, those wild aurochsen that with time would eventually disappear, outcompeted by their newly more domestic kin who from that first day on would always have the help of humans. With humans, the aurochs did far better than it might have done otherwise. It succeeded even as other herbivores, predators, and harvested plants around it were becoming extinct.

While a suite of genes associated with smaller bodies and docility came to be favored in aurochsen as they interacted with humans, a unique set of traits also began to be favored in humans themselves. The traits favored were those associated with lactase persistence (the ability to digest milk as an adult). Adult dogs cannot digest milk, nor can adult cows, pigs, monkeys, rats, or any other mammals. Even adult cats, who we so readily give a bowl of cow's milk, appear to lack the gene to effectively break lactose down. Milk is baby food, or at least it is for all mammals except some modern humans. In order to digest milk as adults, humans had to evolve the ability to continue to produce lactase as adults. Lactase is the enzyme that breaks down the lactose in milk and makes it useful. Our ancestors, it is very clear, could not digest milk as adults. Cavemen and -women, if they drank milk, would have had diarrhea and gas. Perhaps they could have obtained some nutrients from the milk, but not many and if they were already sick, milk and its consequent diarrhea might have even made them sicker. Yet today, most peoples of western European descent, which is to say those people who descended from the first cow tenders, can digest milk as adults. In other words, as aurochsen changed genetically during domestication, so did humans. As we did, we became domesticated too. Once we grew dense enough to need to depend on cows, we could not go back to our hunter-gatherer days. Our lifestyles changed, permanently, as did our genes. We were no longer wild.

Several years ago scientists identified the mutation that is associated in Europeans with the ability to drink milk as adults. (The mutation leads to the production of lactase or, more specifically, lactase-phlorizin hydrolase.) With this sequence decoded, they could study when it arose and how quickly it spread among populations once it had arisen. The first answer came more quickly. The mutant form of the gene for digesting milk in adulthood arose 9,000 to 10,000 years ago, just as archaeological evidence and evidence from the genes of cows suggest that humans and aurochsen came together for the first time.[6] Here, in other words, was a genetic artifact left to tell the story of the earliest days of humans and cows, a marker whose presence in a person would speak to the story of their descent.

Just how fast the gene spread took longer to discern. It required more time and technology, but the question eventually yielded to scientific persistence, more blood samples, and hours in the lab looking over computer printouts of the As, Ts, Cs, and Gs of the genetic code. The short answer was "quickly." In the tribes of our ancestors where cows were first domesticated, those individuals— the majority—who could not digest milk as adults suffered and died. The few who could digest it suffered less and survived. Just as Leigh Binford had long ago argued about agriculture more generally, the dawn of our cattle farming was not some happy bursting forth of innovation, but a more difficult and permanent contract. One might then see that first man who extracted milk from a cow as a kind of early hero, evidence of the way that even in the worst of situations, humans will rise and persevere. Because of his lineage, we abandoned our old mutualists, the many kinds of wild plants and animals we once collected, for the fewer we could grow and farm. As we did, we become dependent on our new partners—our crops. We would continue to change, but only inasmuch as we would replace one species with another, wheat for cows, or sorghum for wheat. Since that moment, we have survived by dint of our crops and cattle, and by them alone.

We tend to think of ourselves as special, and so we call those circumstances in which we come to depend on another species domestication. In truth, though, so long as we and our domesticates both produce more offspring together than we would apart, this is simply a new form of mutualism, one that like our other new interactions is simplified relative to the way we once lived. Humans with the genes for digesting lactose as adults fared better, and so did aurochsen with genes for being a little kinder to humans, for mating in captivity, and for producing more milk. Because of humans, the aurochsen with those genes were able to eat more grass, something they would never have found on their own. Although humans could not, without the aurochsen as intermediary, eat grass, they could make more grass by burning and cutting down forests. They could also reduce competition by killing the other animals that ate grass. Out of that first interaction came a mutual dependency. We depended and would continue to depend on the aurochsen to produce enough food to support the great densities to which our populations would rise. The aurochsen would depend on us to make ever more grass and to kill everything that might compete with or kill them in those new grasslands. Together we began to remake the world, not because we could, but because we had to. Once we entered this relationship, there was no turning back. With our help, the aurochsen that became cows outcompeted those that did not. Wild aurochsen are now extinct, as are nearly all of the big herbivores with which they once competed. We killed those competitors for the aurochsen and killed their predators too. Nor was it just the aurochsen's competitors that would disappear. So too have most of the gatherers around the world, peoples pushed farther and farther into forests that are, each year, smaller and smaller, forests that have given way, by and large, to our aurochsen and our crops, the new mutualists.

Today, there are more than a billion descendants of aurochsen (cows) on Earth. We outnumber them, but we do not outweigh them, and so, depending on how you count, it is ambiguous whether they or we have had, in our coming together, more success. This reckoning is not quite right though, because the number that should be counted is not all humans and human lineages, but those specific lineages of humans with which cows have coevolved. That was long considered to be just Europeans. That was, it would turn out, wrong. Once again, studying the evolution of our differences would change our understanding not just of who we were but who we are.

Roughly ten years ago, the geneticist Sarah Tishkoff, now at the University of Pennsylvania, began to wonder how it was that the Masai, in East Africa, who tend cows by the thousand, drink milk. Tishkoff knew the story of Europeans and the aurochs. She also knew that in addition to Europeans, many other peoples around the world drink milk. The Masai and other cow-tending tribes spent and spend their days following their herds from place to place. In doing so, they drink great quantities of milk, as do many other pastoralists across Africa. In addition, it is known that cattle were domesticated independently in Africa (probably in the northeast) and spread south from that point of domestication. They appear to have spread to western and southern Africa and with them developed intricate cultures associated with cows and milk, cultures like that of the Masai. But the Masai and other tribes are obviously not the descendants of the Europeans who domesticated cows, and so it is difficult to understand how they could possibly have the gene necessary for drinking milk as adults.

Maybe, one might think, the Masai did something special to the milk to make it more digestible. Producing cheese out of milk reduces its lactose content, for example. But the Masai and other East African groups were not making cheese. For a while, it seemed possible that there had been migration of individuals (or their genes) from Europe into parts of eastern and western Africa where pastoralism was found. Then, a group of scientists in Europe found some of the genetic variants apparently associated with lactase persistence. When they checked for the presence of those variants in different populations, they found that they were, amazingly, present both in European descendants of dairy people and in the Fulani and Hausa, West African pastoralists. Apparently migration had occurred that allowed the mutant European gene variants for drinking milk as an adult to move across Africa, or at least to the Fulani and Hausa. But this result did not sit totally right with Tishkoff. It seemed like something was missing and so she collected more data on other pastoralists, which is when the problem arose: the Masai and the Dinka and other groups in eastern Africa lacked the gene variants for digesting milk as adults.

The Masai did not process their milk, nor did they have the European gene for digesting milk. Tishkoff decided to consider a third possibility, that these groups had, in their long history with cows, evolved the ability to digest milk independently of Europeans, that they had relived the history that Leigh Binford imagined was common in the origin of agriculture, but even more than that, that they had done it with exactly the same species. Perhaps individuals with mutant variants of genes that allowed their lactase to persist into adulthood had independently been favored separately in eastern Africa and Europe. It was a long shot, but Tishkoff went ahead.

The answer Tishkoff found in East Africa was that the ability of human adults to digest lactase evolved more than once. It evolved once in Europeans, around 9,000 to 10,000 years ago, at about the time that archaeological evidence and cow genes point to the domestication of cows in Europe. It then evolved again, at least three times, in Africa, beginning around 7,000 years ago, again just about when evidence suggests cows were domesticated for the second time. At least twice (and probably more like four times) upon a time, aurochsen were domesticated. In each of these cases, Tishkoff has shown that individual humans who had the genes to digest milk as adults had far more children who survived to have children themselves than those who did not, and so on into subsequent generations. Their family tree grew branches and with them the genes that allowed them to drink from the land spread quickly all over Europe and Africa. These were the biggest genetic changes in our recent histories, at least that we know of so far. They were repeated, just as Binford had predicted, with nearly identical fates. In both cases, individuals who could not reap the benefits of the domestication of cows died or simply failed to reproduce. Milk did the body good because these were hungry times, times and places when extra nutrition and extra liquid made the difference between passing one's genes on and not. With milk, much, but not all, of human population changed.

In the end, the story of cows and humans is an example of an even bigger, broader story. Wheat would save us, just as the aurochs did, and as would manioc, rice, and our other staples. As our peoples grew denser, we would find ourselves repeatedly at the point of starving and being saved. As the origins of our other crops and animals and the peoples who tamed them are studied in more detail, it seems likely that in many, perhaps most, cases, it will be shown that the genes of those who began to farm changed. It is already known that people who live in regions where grains were domesticated have extra amylase genes, amylase being one of the enzymes that helps to break down starch. Whether these genes too swept quickly among populations has not yet been studied. It seems possible. In fact, it seems possible that in each place that agriculture arose, our bodies changed, independently and differently. Our great human variety reflects, in no small part, the great variety of ways in which we came to depend on individual species, a new less diverse set of species, to make it through the toughest years.

In the villages of our descent, we turned to these new species and latched on, the way a baby first latches on to its mother. We had been brave and independent, but in those moments, we gave in. We would live, for each day after, where and how those species needed us to live in order to benefit from what they offered. We made an evolutionary contract from which we have never since been separated. It is far easier to divorce your spouse than to divorce agriculture. Of course, one can go off the grid and hunt and gather for food, but it is no longer easy and we can't do it collectively, not as a species or even as a country. There are not many places to go, and we have forgotten how to live in those wild ways. The same is true for our domestics too. Our dogs can turn feral, but they never go far. They depend only on us, whereas we now collectively depend on many species, but even this is deceptive. By some estimates, 75 percent of all the food consumed in the world comes from just six plants and one animal. If cows went extinct tomorrow, millions of humans would die, just as would happen with wheat or corn and as once did happen with the blight of potatoes. Cows may look at us with mopey-eyed stares, but we are partnered. Whatever our fate is as we move forward, it is largely shared.

What are not shared are our genes. The genes you or I have today were shaped by those events in recent history when, as a function of our new cultures and ways of surviving, some individuals passed on their genes and others did not. These differences persist. It is hard not to wonder who you are and how your own history, genes, and even (to revisit earlier chapters) microbes influence your modern fate. It is hard, particularly given that despite our differences, we are now converging on similar "modern" diets and lifestyles. The milky, high-fat, high-salt, high-sugar lifestyle we are imposing on our variety of different stories and genes has consequences for our modern health that depend both on who you are and also on who your ancestors were. Even today, it matters whether your people were the ones who crawled under a primitive cow or the few others who looked on from a distance and took the time to point and laugh.

8

So Who Cares If Your Ancestors Sucked Milk from Aurochsen?

A billion people on Earth are overweight, their stomachs pushing at their waistbands and their bodies burdened by excess. The situation is at its worst in the United States, but other countries are quickly catching up. Yet even in the United States, not everyone is affected. Sixty-five percent of U.S. adults are overweight, but the others are not.[1] It is easy to attribute this variety to differences in diet and exercise, but lifestyle is just part of the story. There is also, embedded in our differences, a kind of mystery. After all, the vast majority of Westerners now eat diets that are derived from relatively few domesticated animals and plants. Perhaps you eat only grapefruit or organic camel's milk. Perhaps you practice moderation and make thoughtful, conscious decisions. If so, you are an exception. The average American and, increasingly, the average Westerner has a diet in which three-fourths of all calories come from dairy, cereals (grass seeds), simple sugars, vegetable oil, or booze.* None of these foods was consumed before the advent of agriculture. Ten thousand years ago, tens of thousands of plants would have been harvested from the wild by humans. Agriculture, even in its earliest incarnations, decreased both the total number of foods we harvested as a species and the number of kinds of foods

---

*A list that if mixed, produces nearly all of our favorite foods, from cookies to cakes to cereal, pizza, muffins, pretzels, ice cream, and nearly all the rest.

that any individual human was exposed to. With time, more crops were domesticated, and we recovered some of the diversity of our foodstuffs. But since then, we have come to focus on those few species that grow best and that most simply suit our taste buds.[2] In the process, we have both neglected many crop species (several crops are extinct and nearly a thousand are regarded as endangered) and forgotten how to collect the foods we once gathered. Wild berries now sit unpicked on their stems and a small handful of crops constitutes most of the calories we consume globally. Sure, you can find quinoa in your local Whole Foods, but as a proportion of the calories consumed by humans today, such boutique crops are a tiny drop in an ocean-sized bucket filled with corn, wheat, rice, and a few pieces of crispy meat.

How we metabolize our relatively new diets differs, at least in part, because of the differences in how our recent ancestors lived. Imagine that we performed an experiment in which we fed every person the same foods in the same quantities. We could come back and check on their (or really our) status through time. What do you predict would happen? We tend to act as though everyone would begin to look the same, or at least similar, in terms of their weight and health. This is the premise on which nearly every diet plan, exercise book, and weight-loss show is based, be it the grapefruit diet, the all-meat diet, the no-fat diet, or something else. It is the premise on which growth charts for babies are derived. It is the premise on which most of medicine, in one way or another, relies. The truth is that we would still differ even when eating exactly the same food. Those differences are the result of the differences among our pasts, differences that assert themselves from just beneath the surface like some sea monster faintly visible in the dim light of our collective minds.

Let us return to the story of milk. As I have discussed, we do not all have one of the versions of the gene required for digesting milk as adults. Geographically speaking, the ability to drink milk as an adult remains relatively rare. No Native American populations, whether the Incas, the Maya, or any of the several thousand other groups, were able to drink milk prior to the arrival of Europeans and their genes. Roughly 25 percent of the people on earth are totally unable to digest lactose as adults and another 40 to 50 percent can only partially digest lactose. If these billions of individuals happen to drink milk, they will suffer diarrhea and gain, on average, about 5 percent less weight from the standard American diet than do those who can digest milk. In disease-prone environments, such individuals are also at more risk for dehydration associated with diarrhea. In the context of the villages of our origins, any individual who received 5 percent fewer calories would stand a lower chance of passing on their genes, and so it is that historically the milk-digesting genes won, at least where we domesticated cows. In the context of our modern world in developed countries, in which calories abound, the 5 percent extra are more likely to be chalked up as being to our detriment. So, while the advertisements may say that milk "does the body good," they fail to mention the caveats "but only if your body can digest it" and "only if you need it." That our bodies respond differently to the same food as a consequence of our ancestry may seem obvious. Yet we ignore such realities every day. The USDA food pyramid still has as one of its main items "milk," along with fruits, vegetables, meats, and beans, even though most humans worldwide cannot digest milk. Milk is just the beginning of the unraveling of the idea that any one species of plant or animal food (or processed version thereof) might do us all good.

It is only in light of our recent evolution that the differences among us in our genes make sense. Take your saliva as another example. Amylase is one of several enzymes present in the saliva of many animals, including you. It aids in the breakdown of starches such as those found in corn, potatoes, rice, yams, and other staples of both early agriculture and modern diets. Some humans have extra amylase genes, and so produce more amylase. Those individuals digest starch more quickly and efficiently. Some of us, perhaps you, have sixteen times more amylase than do others. In historical context, this variety appears to exist for a reason. Our preagricultural ancestors had few copies of the amylase genes,[3] and therefore less efficient spit. But it seems that in those peoples that began to farm starchy foods, individuals with more copies of the amylase gene did better, and so passed on their genes. In the parts of the world where humans still regularly starve and eat starchy foods, such as much of the developing world but also poor regions of the developed world, having extra amylase genes may still be beneficial. It may allow some individuals to garner more energy than others from the same bag of rice. In those parts of the world where we eat too much, the same genes are likely to help to make us fat. How our bodies respond to the food we give them depends both on the ways in which our recent ancestors lived and on where we now live. One man's survival gene is another's belly roll.

Nor are the descendants of agriculturalists (whether they were dairy or potato farmers) the only ones likely to have unique genes. The descendants of hunter-gatherers may also have specialized genes associated with their former lifestyles. Ten thousand years ago, all of our ancestors were hunter-gatherers, though they gathered different things as a consequence of where and how they lived. Some populations ate mostly meat, others mostly insects, and still others diets rich with bark (I did not say tasty, just rich). By 5,000 years ago, many fewer peoples survived by gathering and hunting alone. By 1,000 years ago, there were fewer still, and they were located mainly in marginal climates, seasonally cold or dry places like deserts and the Arctic, where crops did not grow well enough to supplant gathering. In general, one would expect that at least some of those peoples that persisted in hunter-gatherer groups might have evolved genes associated with the unique challenges of their ways of living. This should be most likely in those groups at the margins of survival, in places so inhospitable to thriving that the agriculturalists did not ever bother to push them out. In those margins, the genes that were useful are likely to have been very different from those that were useful in dense agricultural societies. But how?

James Neel, an anthropologist at the University of Michigan, suggested the hypothesis that hunter-gatherers who lived in seasonal environments, where food is plentiful sometimes but scarce other times, would have had an advantage if they could store food on their bodies (get fat) more quickly during the good times.[4] This idea, often called thrifty genotype hypothesis, has precedent in ecology, though no one seems to have noticed.

In the early 1800s, the German doctor Karl Georg Lucas Christian Bergmann argued that animals that live in cold places tend to be bigger and fatter than those that live in warm places. His explanation for this pattern was that big, fat animals have less surface area relative to their volume (a snake is all surface, an elephant all innards, for example) and so they are more likely to stay warm enough in the winter to keep from freezing. Over time, Bergmann's original rule was broadened when it became apparent that there were two reasons to be fat where it was cold: to stay warm (as Bergmann originally suggested) and to keep from starving during the months when food was too scarce and a period of fasting was inevitable. In other words, fat was useful both where it was cold and where food was seasonal. Fat is a meal when there is otherwise none. Societies, though, have other options. They can, like honeybees, store food in the winter. The question becomes whether humans are more like grizzlies or bees.

Although often discussed, Neel's hypothesis has been poorly studied. He may be wrong. To test Neel's theory, one could look for genes that tend to be favored among hunter-gatherers, though they might be different in different hunter-gatherer groups (much as different mutations in Africans and Europeans allow adults to drink milk). Alternatively, one could check to see whether hunter-gatherers, when confronted with modern agricultural diets, tend to suffer disproportionately from diabetes and obesity, as might be expected if they were especially efficient at converting food to fat and simple, easily used sugars. Neel's theory predicts that hunter-gatherers from seasonal environments should produce more sugar and fat from this modern diet than do individuals either from hunter-gatherer lineages from less seasonal environments (such as tropical wet forests) or from agricultural societies. In other words, they should be more likely to be obese (good fat storage) and diabetic (really good at turning foods into sugars). One would predict that diabetes should be common in the peoples who lived in those same places where ants and bees store honey or bears get fat, which is to say in deserts, subtropical lands, and tundra.

In 2007, a large study compared the prevalence of diabetes in hunter-gatherers and former gatherers to those of human populations more generally.[5] Many hunter-gatherers store food, particularly those in the north where seal and salmon can be dried. In those groups, diabetes is still rare, actually more rare, for example, than in Western societies. In contrast, in desert and subtropical groups, whether in Australia, Africa, Asia, or the Americas, where storing food is more difficult, diabetes is four times more common than in agricultural societies and Western societies. Only two possible explanations exist for this pattern. Societal pressures may push hunter-gatherer populations consistently and nearly inevitably toward lifestyles that are poorer in nutrients and simpler in sugars than the general population. Alternatively, though not mutually exclusively, hunter-gatherers suffer today because of the consequences of their once uniquely useful, but now misplaced, genes.

No one yet understands how all the genes for metabolism merge to create the different responses to similar diets seen among different peoples. The genes for bodily processes like storing fat are far more complicated (and ancient) than are those for drinking milk as an adult. Whatever the answer, it is unlikely to be simple. It is also unlikely to be the same from one gathering group to another. At the same time, it seems predictable that as humans spread around the world, their metabolisms changed to reflect differences in diet and lifestyle. The more we look, the more we are likely to find our differences. Those differences are due to our many histories and they bear consequences.

Clearly, our ancestors' diets influence how our bodies respond to our current diet. More genes will be discovered that relate to our distinct histories. Some of the genes associated with how we treat each other, our social behavior, for example, seem to differ dramatically from one group to another. Perhaps, it has been suggested, human lineages that moved into agriculture required different social abilities than those that did not. Perhaps they had to be more docile, less prone to aggression. Perhaps they had to be more like their mopey-eyed cows. Perhaps. A dozen labs are studying this question, so we may know soon. In the meantime, we continue to walk around, each a little different from the other. Some of those differences are pure chance, the wonder and idiosyncrasy of inheritance. Others, like genes for digesting milk or breaking down starch, are adaptations for who and how we once were.

In the end, we are different because "we are each special flowers" as our preschool teacher might have told us. We are special flowers because over the last 10,000 or so years of our evolution, we have lived many different lifestyles in many different ecological contexts. Once we all ate the wild species that were around us. Your malaria is my dairy cow. Your fasting period is my extended history of poor-quality grains. Western medicine does not consider these differences, perhaps in part because of a mistaken assumption about our variety, yet another assumption that Sarah Tishkoff would correct.

In standard evolutionary trees of humans, there is a thin and poorly leafed branch relegated to Africans; other branches support Europe and Asia. From those branches split off the twigs leading to Asia, Australia, and the Americas. What Tishkoff came to realize was that something was missing in this picture. Although modern humans had lived in Africa nearly twice as long as anywhere else, Africans were poorly included in studies of the tree of human life.

They were an afterthought, a couple of guys someone had taken a DNA sample from while on a trip. Yet it has long been known that Africa is incredibly diverse culturally, and so it might have been suspected that it would also be diverse genetically. Tishkoff's work has now shown that it is. In fact, by some measures, the genetic diversity among African groups is as great as that found in all of the rest of the world combined, a finding that reconciles with what is known about the diversity of cultures. Almost a third of the languages are found in Africa and, with them, a third of all ways of living. In other words, in the most common telling, we had the tree of human life precisely backward. The tree of life itself is rooted in Africa, where most of the branches remain. The rest of humans, from Native Americans to aboriginal Australians to Swedes, descend from just a few branches corresponding to one or two migrations out of Africa, and with those migrations came a loss in genetic diversity. A smaller subset of people and genes moved over each further hill.[6]

There are consequences to the realization that most of human diversity can be found in Africa. For one, it means that our categories of black, white, and brown are useless when thinking about health and disease. The white patients that were long the focus of much of Western medicine come from relatively few, and somewhat anomalous, branches of the human tree. In medicine, where nonwhite races are considered, it tends to be as a contrast. For example, the inability to digest milk as an adult was long thought of as a deficiency. Yet it is the normal condition. If anything, having the ability to drink milk as an adult is unusual, both in our history and in most parts of the world today. Similarly, thousands of studies compare the differences in health and disease between whites and blacks or whites and Asians with a kind of implicit model of trying to understand diversity. Such studies almost invariably find differences that are then attributed to genes, culture, economics, or some complex slurry thereof. But the truth of the matter is that in any such comparison, the artificial category "black" or "Asian" includes much more variation than the category "white."[7] Whiteness, after all, is as unusual as the ability to drink milk as an adult. To the extent that genes influence our modern differences in health and well-being (and they clearly do), this racial split will remain "of limited utility," or, to mince fewer words, stupid. What we really need is a kind of evolutionary medicine that acknowledges the diversity and contexts of our pasts.

Practicing medicine in a way that acknowledges the differences in our histories is difficult. It depends on knowledge of our many histories, a knowledge that we are losing rather than gaining. Once upon a time, which is to say, 200 years ago, there were, by most estimates, around 20,000 cultures or language groups on Earth. Not all of those cultures were associated with specific genetic adaptations to particular mutualists (or for that matter, worms, pathogens, or climates), but many were. As of today, there are just 6,000 or 7,000 surviving language groups and associated cultures. Perhaps 1,000 or fewer of those are viable in such a way that they are likely to endure another two decades. Each day we lose a few more stories encoded in languages, histories, and spoken and unspoken words.

We have swapped our old partners, wild species of fruits, nuts, and prey, for our new domesticated partners. The relatively new lifestyle associated with these new partners continues to spread, often without its associated genes. In the last 200 years, a few agricultural cultures have grown ever closer to replacing the other remaining cultures on Earth. As this happens, the relationships between our individual histories and our cultures grow muddled. In the future, few individuals are likely to know enough about the livelihoods of their own ancestors to be able to make sense of their genes. Maybe you are already one of them, too disconnected from your past to know whether your ancestors dined daily on seals or tended instead toward the delicacy of wild fruit and ant queens. Soon, few people will be able to say much about where they came from and what they once did. Yet the genes and descendants of these cultures and peoples will persist—rolled into our broader genetic pool and with it our wonderful, tragic, muddy story. When genes are still linked to the cultures in which they evolved, one can understand their origins and significance. But as cultures homogenize, our individual histories blur. Today, Masai individuals can be identified as a cattle people. In that context, their genes for digesting milk make sense. But for the Cavineños, a very small indigenous group in the Bolivian Amazon, the history is less clear. They now farm, somewhat, but for how long have they done so? What did they farm in the past? Some individual Cavineños still know, but most do not, and therefore that knowledge will soon be lost, and they will be culturally like other Bolivians living in the wet tropics. Their genes, though, will be otherwise, a more complicated story that asserts itself without context, a glacial rock in the middle of a field around which one must walk and plow, but which, on its own, never quite makes sense.

PART V

How Predators Left Us Scared, Pathos-ridden, and Covered in Goose Bumps

9

We Were Hunted, Which Is Why All of Us Are Afraid Some of the Time and Some of Us Are Afraid All of the Time

Our parasites and mutualists influenced our bodies. It is the predators, though, that messed with our minds. We come from a long line of prey; we have been eaten since we were fish. For most of our history, we were more like the pronghorn than the cheetah, more likely to flee than to chase. So it is that time and natural selection had, until very recently, favored the wary over the brave. You can experience your body's wariness to the threat of a predator when someone jumps out at you from a dark hiding spot. You can feel this past when you watch a scary movie or even by reading about someone else's scary experience, for example the day in 1907 when a girl in India named Bakhul and her girlfriends went to gather the leaves of walnut trees to feed to the cows. Bakhul had climbed high into the branches to reach the tenderest shoots, the cows' favorites.[1]

On that day, she was the first to finish gathering and to start down the tree. As she did, she felt a tug on one of her small feet. Was it one of her girlfriends? No, it was firmer, less playful. A tiger stood at the base of the tree. It looked up at her with big eyes and pawed again with its claws extended. It pulled her as if she were a lamb. She screamed and held on, but only for a moment. Her leaves fell around the tree, as did the beads of her small blue necklace. The tiger carried her into the woods. She screamed. She was terrified, but still alive.

When they were told about the tiger, Bakhul's mother and father were despondent beyond words. One town over, a woman had seen her friend taken by this same tiger and it had left her mute, in shock. Bakhul's parents too seemed unable to speak. The wife stirred the pot of their food. The husband sat, unhinged. The door of his life had swung open and it could not be put back. Somewhere at the edge of town or in the quiet between houses, the tiger was walking. Bakhul might still be alive, but no one chanced searching for her, not yet. The families shook and waited inside their houses for whatever might come. Lightning strikes the same spot only once, but tigers can strike repeatedly. This tiger had already killed more than 200 people in Nepal before armed guards chased her across the border. Once in India, she had killed another 237. Now it was in this town that she would do whatever she did next. Given her history, she would almost undoubtedly eat someone else. If not Bakhul, then who?

In this story, one wants to yell at Bakhul's family to look for her. "Search her out!" "Be brave!" No one would have listened. The town was shuttered in fear. Doors stayed closed. Children peed into cans and poured them out the windows. Adults too searched for their own vessels or crouched just outside the door. This society had turned inward and gone rancid with fear and excrement. No one wanted to leave their houses, even as food began to run low and the crops rotted on their stems. Even baboons, stronger, faster, and better defended than humans, hang close to their kin when predators are near. They sit back-to-back looking out and groom each other, gently touching each other's heads and backs, just as it was in this village, people tending to each other with equal parts of tenderness and alarm.

As the villagers waited, they had time to recount other stories they had heard about this tiger and when those stories were exhausted, about other tigers. One town away, in the village of Champawat, a group of men was walking along a path near the village when they heard screaming. Then they saw a tiger coming toward them, and in its mouth, a naked woman, her long hair dragging as she cried for help. In that story too, the men had been too scared to act and so the tiger, carrying the woman, continued on its way. There were dozens more stories. Most ended fatally, but every so often the person was saved, and so they hoped for Bakhul. They hoped for her to stagger back into town, hoped because they were all, each and every one, too terrified to act.

Bakhul's story has lived on through the writing of Jim Corbett, the great hunter of man-eating animals. It was Corbett who would eventually come to try to find Bakhul and then to kill the tiger. The longer story of humans and predators, though, is embedded in our bodies, in our genes and their products, in particular in a network of ancient cells in our brains called the amygdala. The amygdala is connected to both the more ancient and the more modern parts of our brains. It, along with the adrenal system, is caught halfway between our deep past and the present. From there, it urges us into action or contemplation, depending on the circumstances. If you feel any eagerness on behalf of Bakhul, any modest restlessness for resolution of her story, an interest that perhaps even sent a chill or two down your arms, it was because of your amygdala and its signals. But more generally, it was because you are descended from a long line of individuals who escaped being eaten, at least long enough to mate, a lineage going back not just to grandma but to lizards and then even further. Your heart pounds harder when you are afraid (or angry, a point to which we will return) because of the pulleys and levers of your adrenal glands and the signals sent from your amygdala to your brain stem, that even more primitive root of our actions and wants. This system, sometimes called the fear module, evolved primarily to help us deal with predators, whether by flight or, less often, at least historically, fight, but it's a finicky system that can be aroused at the mere idea of a threat. Fear, or at least the urge that precedes it, may even be our default reaction to our surroundings. Some elements of our amygdalae appear to constantly send out signals to our bodies that we are afraid. Most of the time, other parts of the amygdala suppress those signals. But when we see, hear, or experience something that triggers fear, the suppression is released and fear courses through us, instantly, like a bomb in our brain.

Our fear modules have been shaped by many thousands of generations of killings and near escapes, since the very earliest moments in which one animal pursued another. Now we retaliate against these predators, but for most of our long history of interacting with predators, we did not have guns. We did not even have the wherewithal to pick up and wield sticks. We screamed (the scream itself being a near innate element of the fear module) and ran. If we did not, it was just a matter of time before we went "one by one into the capacious stomach of [our] arch-enemy, which never neglects an opportunity of reducing [our] numbers and thus fulfilling its mission in life."[2]

When asked around campfires and poker tables to construct stories of our identity, we tend to cast ourselves as the predators, both powerful and in control. In storybooks, Little Red Riding Hood is saved, just in the nick of time. That may be who we are now—Johnny to the rescue with his big, bad gun. But the truth is that for most of our history, we were not able to save the little girl in the den of the wolf. We may have tried, but more often than we succeeded, we failed, at least for the first several hundred million years. By the time Bakhul was attacked, such attacks had become less common and yet still occurred (and occur). The tiger that had found Bakhul would turn out to be a "man-eater," an individual that, through injury or age, had become unable to attack its ordinary prey that might fight back. For most of human history, though, our ancestors would have been eaten simply as one of a variety of possible kinds of prey. As we grew more common, we may have actually become preferred. The more of us there were, the easier we were to find.

Predators did not avoid our ancestors until they had weapons. Even then, the entire concept of man-eaters suggests our ancestral weakness rather than our strength. Injured and old "man-eaters" eat humans because we are the easiest to catch and kill. We have no horns, sharp teeth, or even hair to impede digestion. We are nearly as well packaged for consumption as a hot dog. The "Man-Eaters of Tsavo"[3] are rumored to have killed dozens of people in Kenya, enough to impede the construction of the railway from the shores of Lake Victoria to the port of Mombasa in the late 1800s. The two male lions were eventually shot and shipped back to London, where they sat ignored in a museum for a hundred years. However, a study of the bones and mouths of these lions shows that they had been sick, in one case deformed and missing a number of its teeth. In other words, if you are an old predator that cannot catch anything else, a human is your best alternative. We are among the only animals, aside from a three-legged wildebeest or our simple-minded cows, so defenseless that even an animal with a broken leg or missing teeth can catch us. Today, those animals that attack us tend to be killed (as eventually were the Man-Eaters of Tsavo), but for most of our history they operated with impunity. Our ancestors could barely see in the dark, and so when they heard a sound in a cave they crouched, listened, and hoped that if it was a tiger, bear, or other big carnivore, it would eat somebody else first. Imagine peeing in the woods at night with the stars overhead and the sounds of lions or tigers and other animals in the grass all around. It was perhaps with such thoughts in mind that San Bushmen rendered cave paintings of lions dismembering humans.[4] Such scenes have long haunted our dreams.[5]

In the wild history of humans and large predators, we were long unambiguously the prey, which allowed the fear modules in our brains, modules that developed many millions of years ago, to persist and even to become elaborate as we evolved. To find a predator in our ancestry, we may have to go back to the time when we had four feet, a lizard tail, and scales. Even then, we were probably

just as likely to be the eaten as the eater. We have been screaming the barbaric, animal equivalent of "Oh shit, don't eat me" for
300 million years. Four kinds of data tell us that we were eaten up
until very recently. First, large numbers of actual predation events
on humans have been recorded. In colonial India, tigers may have
eaten more than 15,000 humans a year.[6] At least 563 people were
killed between 1990 and 2004 by lions in Tanzania alone. Nor is
it just tigers and lions. Mountain lions eat people. Giant eagles eat
(or at least ate) children. Bears of quite a few varieties eat people.
Lions, leopards, alligators, crocodiles, sharks, and even snakes eat
people, especially children. Even wolves eat a runner or two. And
all of this is just during recent years when predators are both more
rare and far less diverse than they were during most of our evolutionary past.

Second, the fossil record of our kind is rife with the bones of
our frequently gruesome demises. An individual of *Australopithecus
africanus* was discovered with eagle talon marks in its head. It was
discovered in a pile of other bones, under the eagle's nest. In a study
of Pleistocene leopard diets, *A. africanus* individuals were the most
common prey item at one site, which is to say the leopards were actually specializing in our ancestors. At a second site, another leopard bone pile revealed similar results. At both sites, the bones that
were left to be found by scientists were those of the head, which
leopards do not eat and then, in a pile, the regurgitated pieces of
the rest of the body.[7] One imagines our ancestors living together in
a small group that was, night by night, victim not just of leopards
but also of the other predators that walked among shadows, invisible to our crude senses,[8] predators like lions, hyenas, and wild dogs,
their larger, now-extinct kin.[9] Nor are these south African caves an
exception. Among the earliest fossils of hominids, many come from
bones that appear to have been broken to bits by carnivores. We see
our own history, our mammalian history, most clearly through the
mouths of big dogs and cats.

But the most detailed evidence of the role of predators in

shaping our identity comes from other primates. For most of our primate history, we would have been about the size of a capuchin monkey and so suffered, many days, a fate like its fate. Several species of New World eagles, such as the Harpy, eat monkeys preferentially. Leopards sneak into trees and take monkeys. A recent study in the Ivory Coast followed two individual leopards and found that although the two individuals had different dietary preferences (one cat's pangolin is another's giant rat), almost half of the diet of both was primates, including big monkeys and even chimpanzees.[10] In fact, in the few places where primates have been studied over many years and where large predators are still relatively common (if not as common as they once might have been), more individuals die in the mouths of predators or from snakebite than to nearly any other cause. Baboons have been particularly well studied (and eaten), and in those studies one finds evidence of predation by eagles, hyenas, wild dogs, lions, leopards, jackals, cheetahs, and even chimpanzees.

Where predators are common, perhaps three out of a hundred monkeys (or apes for that matter) die each year by being eaten, as was likely our fate for much of our history. For context, in a given year, cancer kills only one out of a thousand Americans, which is to say that if early humans were like modern primates, death by predation would have been 30 times more likely at any moment than is death by cancer today. More to the point, cancer tends to kill us after we have reproduced, whereas predators knew no such forbearance. Regardless of when humans escaped predation by making better tools or being smarter, we began like the other primates, eaten often and nearly inevitably.* To the extent that we were dif-

---

*I know, I said there were four kinds of evidence. The fourth form of evidence is the most equivocal, but also in a way the most interesting. *Toxoplasmosis gondii* is an incredibly common disease in modern humans. In most adults, it is benign, dormant even, but for the fetuses of pregnant women, it can be deadly. The funny thing about *T. gondii* is that it is a cat parasite. It cannot complete its life cycle unless it moves from a human (or whatever other temporary host) back into a cat. One wonders why *T. gondii* bothers infecting humans. Perhaps it is a mistake. Many of us live with cats now, and so pick up *T. gondii* accidentally, and it does not and has never gained from the interaction. But a second possibility exists, the one I prefer. It is possible

ferent, it seems possible that we were more rather than less edible. We were potentially easier to track than were other primates, because of our heavy footprints and, at least according to one anthropologist, smellier bodies. Those few primate species, such as vervet monkeys, that have calls with specific meanings, almost inevitably have calls that relate to the threat of predation. The vervet monkeys have three words, "leopard," "eagle," and "snake," which were, in all likelihood among our first words, the most important nouns. Close behind them, one suspects, was the verb "RUN."

Humans, like other primates, were long the hunted, a situation that shaped how Bakhul and her friends responded but also how they lived and, for that matter, how we respond and live. When predators are around, going to the bathroom and sleeping are among the easiest ways to die (particularly since many primates, not just humans, appear to snore). Some of our responses to such threats relate to specifics of our behaviors, such as the way we (as primates) sleep and build our homes. Monkeys and apes build nests high in trees and sleep together, so that at least one individual is always awake and might alert others of a threat. Chimpanzees build nests, typically above three meters, which is perhaps not coincidentally a little higher than leopards can jump. The only earthbound exception, aside from us, is the gorilla, which when it moved to the ground became big and strong, perhaps as a defense against predators.[11] If you cannot climb, you had better be big enough to deal with a leopard on your back, literally.

When we moved to the ground, we were not big enough. As a consequence, we became even easier to eat than we had been before. We may have compensated by moving into caves (as baboons do today) and then, eventually, building houses from which predators could be excluded, houses that we tended to circle, like wag-

that *T. gondii* moves into humans hoping that we will be eaten by a cat, so predictable was that fate in our long history. Maybe. Maybe not. It seems telling that *T. gondii* is not the only disease with this life history that infects us. There are several, all waiting for a tiger to eat their human, so that they can mature.

ons, when predators were near, doors facing inward. Doors were small and defensible and we nearly always lived in relatively sizable groups of ten or more, even when such groups required us to walk long distances to find food.* The Mbuti pygmies once built huts reinforced like a cage, although it was to keep animals out rather than in. In Bakhul's town, the houses were clustered, as they may be in yours. The gated subdivision and its cul-de-sacs is a modern version of our earlier villages, in which our front doors face each other and we all keep an eye out for what might lurk in our collective shadows, even though such a design is inefficient and, for a variety of reasons, actually more dangerous than a grid of streets. We feel safest this way because once upon a time we were. So it is that we lock our doors each night and that, in Bakhul's village, they boarded theirs.

Predators also influence, even today, when we do the things we do. Humans and other primates do very little at night. We sleep in groups but are otherwise inactive, for good reason. Our senses are dulled to night's realities and dangers. One of the very few things we do at night is give birth. In those few places where human birth is not induced, like Bakhul's village, most babies are born in the dark hours between dusk and dawn. A recent study of zoo chimpanzees found that roughly nine out of ten were born in the middle of the night, not long after midnight.[12] If you are older than fifty, odds are you too were born around 2:00 AM. Being born in the middle of the night, when your kin are gathered around, sleeping and, if the need arises, defending, may decrease the odds of the mother and baby being eaten in the process of giving birth.

My wife and I once stumbled upon a black-and-white colobus monkey in the hollow of a tree. In her arms was a pale white newborn baby. It looked fragile and defenseless, like our own newborn

---

*One wonders whether we really had a choice. Could we have constructed another kind of house? Yes. For example, consider the houses of birds who build most of their nests with open tops. They go about, unworried by the rain. Some birds build coverings over nests or build nests in hollow chambers, but they are more the exception than the rule.

daughter and son were. I cannot imagine trying to flee a predator right after my wife gave birth. I suspect all we would have been able to marshal would have been something along the lines of what we told the nurses, "just give us some time." Newborns and new mothers (and at least this new father) need all the help they can get. Perhaps tellingly, the only monkey known to give birth during the day is the patas monkey. Patas individuals are together during the day but apart at night. These patterns in birthing may or may not be linked to predation. So far, no other possibilities have been suggested.

The effects of predators on when we give birth and how we build our homes are still a bit speculative. But there are also less ambiguous consequences of having been eaten across most of our generations, for example, our fear modules themselves, built as they are of elements of hormones, blood, adrenal glands, and brain. When the tiger pulled at Bakhul's foot, a predictable series of things happened in her body and, importantly, in the bodies of her friends, who were watching nearby. The cells in their adrenal glands would have released bursts of adrenaline from specialized "packets." The adrenaline would have triggered a chain of other chemicals that would have caused their small hearts to beat faster and with more force. Blood flow would have increased, the tracheoles dilated and lungs expanded, allowing more oxygen into the blood. All of this would have triggered a rush of superpowered energy and awareness and, secondarily, a sensation of fear, a sensation meant to trigger a later, more considered, response. For Bakhul, none of this was enough to keep her out of the tiger's mouth. It still might have saved her from death, though the odds were low. Once tigers catch their prey, they seldom fail to kill it. Bakhul's friends, on the other hand, escaped, in no small part because of the response of their adrenal systems, which evolved specifically to help us get away from predators or, more rarely, to stay and fight.

In wild primates, when the fear response is triggered, an alarm is often sounded, whether something specific like the vervet monkey version of "leopard," or a more general scream. Next, the monkeys usually flee, which is probably the most common response. More rarely, when the predator seems weak or there is no other choice, primates will mob and charge it (though usually from a safe distance—better to not tempt fate or leopards too much). Sometimes such mobs are successful, whether in simply chasing away the predator or even killing it. Other times, they are not. When it is an option, the best route remains to, like Bakhul's friends, run away.

Primates like us are not unique, of course, in having adrenal systems and associated defensive behaviors. Our adrenal system evolved hundreds of millions of years ago. Its basic functioning has remained unchanged across generations, elaborated upon but not replaced. Nearly all vertebrates respond to threats with the same bodily response we experience. What differs from one animal to another is both how the response is orchestrated and how it is fine-tuned. In reptiles, there is no amygdala and so fear runs directly from perception to the brain stem—perception to barely modulated action. In mammals, the amygdala takes over orchestration. It signals the brain stem. It notifies the conscious brain with a signal that we experience as fear. Among mammals, the twitchiness of the amygdalae of different species differs in their tuning. A cow's flight response is relatively dulled to external triggers (though it can still be triggered with enough of a stimulus, as it often seems to be on big industrialized farms). This is one of the reasons cows and many other domesticated animals, such as lambs and even genetically modified salmon, are so susceptible to predators.[13] Cows and lambs are not just meek. They are actually numbed to the dangers that once haunted them, too tame to flee even when the wolf or the butcher is at the door.

Much of the fine-tuning of the adrenal system from species to species, or from one individual human to another, occurs via changes in the abundance of a single protein. On average, humans have a lot of it, as do nearly all but the very largest, most well-defended species. Cows may have once have had much of it too, but we bred it out of them, as we did with many other domestic animals.[*][14] Such changes, while not inevitable (horses are still very twitchy), can occur very rapidly. In a Russian experiment that began in the 1960s to domesticate foxes, animals that were friendly toward humans were produced in just three generations of selection. After thirty-five generations, the foxes were not only friendly but also submissive. They wagged their tails and licked their domesticators' fingers. These offspring had a reduced fear response relative to their ancestors and lower levels of the hormones associated with fear. Evidence suggests that similar transitions occurred in wolves as they made the transition from solitary hunting to hunting in packs (in which twitchiness and rushes of adrenaline and aggression might have been counterproductive to team efforts), with social animals like wolves being docile relative to each other, much in the way that cows are docile to humans.

One could argue that it would be useful for society if humans evolved a less temperamental fight-or-flight response. For the most part though, we appear to have maintained our high responsiveness. If anything, what changed over the most recent years of human evolution is that we became less likely to flee but more likely, with our new tools and bigger brains, to stay and fight. We would even begin to search out fights, as was the case for Jim Corbett, the hunter who was called to Bakhul's village to find her, the tiger that attacked and maybe killed her, and rescue the villagers from their ancient and overwhelming fear. Corbett would find his quarry (though killing it was another thing entirely), and as a species we would find big animals everywhere, chase them, stab them, tire them, and eat them, one at a time.

---

*Though some fear is hard to shake. Domestic hens, though they may still have things to fear, no longer face much threat from hawks. Yet when a plastic hawk is made to fly above tame chickens and wild fowl, both stop eating and walk alert, alarmed. They recognized the hawk. And so the chicken, like us, has apparently retained some of its specific fears even though in its most common modern condition (in cubicles free of predators and stocked with near infinite food), such fears are misplaced. They are more like us than we might admit, these chickens.

10

From Flight to Fight

In the case of Bakhul, the villagers would chase back, but only when Jim Corbett came to town. Corbett was a young man, a boy really, but he came to town to kill a tiger. He wanted to give the village back its peace. With age, Corbett would become the greatest hunter of man-eating cats, but not yet. At this point in his life, he had the wisdom of a young man combined with the bravura of a big gun. With the gun in hand, he turned his fear to rage, flight to fight. He wanted to pursue the tiger, and, once in town, it did not take long for Corbett to find the tracks. At the place near the oak trees, he found blood and the beads of Bakhul's necklace, and began to follow the fresh trail. He had not walked very far, fear and then also rage welling up in his throat, when he found one of the girl's legs beside the river in a pool of blood. Bakhul was very definitely dead.

Corbett bent to her leg, tenderly. As he stood there bowed and saddened, he began to feel as though he had made a mistake. The feeling he had was very specific. The hairs on his arms rose. His skin tingled. A chill ran through him, and he had an overwhelming urge to run. Each of these responses was innate and unconscious, triggered by the sound of falling dirt that he heard coming from the hill above him. It was the tiger, looking down, its heavy paws crumbing the hill's loose soil. Corbett's body released adrenaline, which flooded his blood and he began to feel, for lack of a more perfect word, explosive. His heart leaped at his ribs as though trying to find a way out. Some part of his body was sure that he, like Bakhul, was about to die, and that part of him was urging him, imploring him, to run.

The tiger turned from Corbett and moved on, up the hill. Fighting to control his body's urgings, Corbett got a better grip on his gun. He went after the tiger and as he did, he was suddenly aware of thousands of sounds. He could hear leaves bending back and forth in the wind, insects moving, and then as he got closer, the tiger's low snarl. He was alert with adrenaline. Corbett followed the pad of footfalls and a kind of pounding that might have been his heart, or maybe even hers. He pursued all afternoon until his mind was overwhelmed with the accumulated hormones and blood. He could not make out where he was relative to the village or the tiger, and so he started to panic. He ran over rocks, through blackberry thorns and giant ferns until the sky was dark blue, then black. He had no light, so although the tiger could undoubtedly see him, he could not see her. He was in a narrow crevasse, closed on one end. He had a gun, but he was in this moment as naked and vulnerable as any human had ever been. He could hear her now, again. He backed away, slowly, out of the forest, retracing his steps and listening for the tiger he could no longer hear over his breathing, the tiger that he was sure would follow him home.

Somehow, Corbett was spared and made it back to the village. There, he composed himself. Tomorrow he would try something new, something different. Meanwhile, up in the ravine, among the vines, thorns, and hills, the tiger roared.

No one knows when humans began to hunt on a regular basis. Of course, we long hunted insects, snails, and the occasional rodent in a hole. But what about bigger prey? The answer is mixed in among the messy piles of ungulate and human bones excavated across Africa, Europe, and Asia on the basis of which the question is debated vehemently by anthropologists. They wield pens and pencils in hands that once held spears. They brandish them and grumble, but consensus seems far off. What one can say is that before we invented tools, hunting was likely to have been rare and clumsy.

Even once we had tools, they would not have helped much, not at first. For the first 500,000 years of our human history, we had nothing more than stones sharp enough to break the marrow free of scavenged bones. With these first tools, we were like hyenas, though less dangerous and far less effective. Eventually, early stone tools were combined with sticks to produce spears. Spears were combined with running and calling back and forth, to chase, at first, smaller herding animals. This appears to have occurred at about the time that wolves too started to become more social. We would have hunted side by side with wolves during the day, and hidden from them at night. The transition was slow, but progressive. In at least two well-documented study sites, archaeologists have shown that humans moved from eating slow-reproducing, slow-moving prey, such as tortoises and limpets, to faster-moving, faster-reproducing prey, such as rabbits and eventually birds.[1] As the tortoises and limpets became scarce and the rabbits were chased away, the hunting of deer and other larger herbivores became more important. In some places, such as the cold regions where little in the way of plant food was available during the winter, it may even have become necessary.

The shifts toward hunting caused our bodies to change. Look at your hand. As we began to pick up sticks and stones, our hand bones literally evolved to be better able to handle such weapons. The ways you grip a baseball bat or ball are vestiges of the way your ancestors once picked up sticks and stones, respectively. *Homo erectus* or *Ardipithecus ramidus* would have sucked at baseball, as would our even more distant ancestors. The only way individuals with better abilities to grip would have been favored is if holding a stick or a ball improved their chances of survival or mating.[2] In other words, tools and their use were eventually necessary for survival. Our legs became longer and our lungs relatively bigger. We became better at long-distance running. All of these things changed together. We were still weak and fleshy. We were still among the most edible of vertebrates, but we could work together using sticks and calls.

Jim Corbett, in the context of this long history, had an idea. He would use the villagers to chase the tiger from the wide, uphill part of the valley down through the valley to the only opening on its narrow end, where he could shoot it. In doing so, Corbett would be reenacting thousands of earlier hunts in which, like wolves or African wild dogs, humans had called back and forth to each other, even signaled, as they chased their prey. Just a few years earlier, archaeologists had documented a place where Native Americans had once chased a herd of buffalo down a ravine and off a cliff, whereupon they butchered them. That night he traced the plan in his head, as though he were sketching out a cave painting. In the image was the tiger, hundreds of villagers raising their arms, and then one man lower down, with a gun. The tiger in this sketch, like the predators in hundreds of cave paintings, still looked as though it stood a fair chance.

When the villagers went to help Corbett chase the tiger, they did so with bodies like those that had long chased animals. Corbett's plan was that they would gather what they could—rocks, cans, and sticks—and stand at the top of the ravine. When Corbett and the headman signaled from the bottom of the ravine, everyone was to begin beating their instruments, so that they might drum the beast out of the woods down toward Corbett's aimed gun. In acting out this plan, they were reenacting the imperfect transition of humans from hunted to hunters.*

---

*I am reminded of the apparently true story of China's Chairman Mao and the sparrows. Mao did not like the sparrows of China. They bothered him (as did three other "pests," mosquitoes, flies, and rats). They shat on his porch and ate valuable seeds. So he did what other man-god rulers might have done; he had everyone in the country go outside into their backyards and beat pots and pans to make noise so that the sparrows would fly in fear. The banging continued for days as the birds hovered, unable to suppress their fear, until they fell dead by the millions, confirming the power of Mao relative to nature. But nature is politically neutral and so the next year, the locusts, once eaten by the sparrows were more abundant than they had ever been. A

The villagers were positioned. So far, everything had gone according to plan, but as Corbett and the chief walked, the chief grew tired. He was old. He begged Corbett to let him rest and then, before Corbett had a chance to respond, he took a break. Nothing could be more innocent and natural than his wanting to sit, except that the sinking of his body onto a log, like the lowering of a flag, was taken by the villagers as the signal to begin beating. They began making noise with anger and fury and all of the energy they had pent up in their mourning. They wailed on their makeshift drums. Sure enough, the tiger ran from their bravery straight toward the place where Jim Corbett and the headman were going to shoot it.

Corbett and the headman were not yet ready. They were still high on the hill and so had to start running down to outpace a tiger running down through the brush. If Corbett and the chief did not make it to the neck of the valley, the tiger would escape and kill again. They ran, but they were too slow, at least when paced against a tiger. Then it happened. The tiger burst from the forest when Corbett was still 300 yards away. The chief was even farther away. Corbett assessed the situation and paused. The chief assessed the situation and acted with all of the fury of the entire village. He fired and missed. So it was that the tiger turned to run toward the villagers.

There seemed to be only one possible outcome. The tiger would break through the line of villagers, killing one or more in the process, and then run back into the woods, where it could lurk for days or years to come. But none of the villagers knew what had happened. Having heard the chief's gun, they thought the tiger was dead. They were celebrating. No one was even looking downhill when the tiger began to run their way. They were like lambs. Then dumb luck intervened.

---

plague of such magnitude ensued that the fields were quiet except for the sound of chewing. Thousands of people starved. Trying to eradicate nature always bears consequences.

The tiger heard the prematurely celebrating villagers and fled from the sound back toward Corbett and the headman, who by now were ready. Corbett fired and hit the tiger in the shoulder. The tiger seemed unaffected. It turned toward Corbett, raised its haunches and lowered its shoulders the way a house cat will do before pouncing on a shrew. Corbett was the shrew. He was also out of bullets and yelled to the chief to bring down his gun. The tiger was ready to pounce. What happened next happened so fast as to make all retellings wrong in some detail. Somehow, Corbett grabbed the chief's gun and squeezed the trigger.

The shot missed. Fortunately, the damage had already been done. The tiger fell dead from the earlier shot. Meanwhile, the villagers farther up the hill had never stopped celebrating, and so they continued. Corbett had killed the monster of Champawat and was to become the great hunter of man-eaters.

After Corbett killed the tiger, the villagers would have taken the time for a funeral and sent the body of Bakhul down the river so that her story might make it to the Ganges, to mix with the stories of other women in other villages across India and time who had been eaten. Tigers would kill more people in the years to come, but fewer each year. In the last century, we have escaped our many-million-year burden of being prey. The hundreds of thousands of tigers that once covered India became thousands and then hundreds. More tigers may now live in captivity in Texas than wild in all of Asia. The same is true or nearly so for other predators—cheetahs, leopards, lions, cougars, jaguars, even wolves and bears. The process that began those many thousand years ago, when we first started to hunt in packs like wolves, has nearly reached its completion. Large predators still kill people, a few dozen a so or year, but very few and mostly at the borders of wild lands, where we walk out to pee in the woods and reenact our primitive urge and they, in return, reenact theirs.

Yet the past's effects linger in each of our lives and in much of our mental illness, discontent, and perhaps even choices about where and how to live. Because while we killed most of the tigers, most of the wolves, most of the bears, cheetahs, and lions (though not the giant primate-eating eagles, nor the deadly venomous snakes), our bodies' responses to those predators remain—shaped by the effects of having had to run away for so long.[3]

We still have adrenal glands and the amygdala in our brains that translate what we perceive into our body's response. We still have these structures even though we now stand essentially no chance of being eaten or even chased by a predator. Our fear (and its companion, rage), modulated by these parts of our body, led us to kill most of the animals that triggered that fear. But what then do the systems of bells, whistles, and blood vessels in our bodies that evolved to produce fear have left to do?

One place we still see the workings of our fear is in our demand for scary movies and books. Think vampires, Freddy Krueger, and terrible murders on crime shows, each of which terrifies us and as it does, triggers the same chemicals in our bodies once stimulated by the tigers near our villages. We have come to buy the stimuli that trigger our fear response as though to remind ourselves of what it can do, of the way our blood can pump as we flee.

But the other reality of our modern context is that the same system that once produced fear in us now short-circuits in a world in which many of our stimuli come not from direct threats to our body, but from distant threats. We listen to the news and hear about every murder. We think about our budgets and fear their consequences. These diffuse fears trigger the same responses that the tigers once did, but they do so chronically, a little bit each day. Instead of yielding a resolution, this fear yields anxiety and stress. As many as one in three adults will at some point suffer anxiety disorders brought on by misplaced fear. If prolonged, such misplaced fears can even lead to depression and other stress-related diseases, and to shorter lives. Our chronic and misplaced fears create chronic stress and distress, and make us more rather than less likely to die. We can wake up at night ready to run from our unbalanced checkbooks as often as we like and we will never quite get away. Our anger, in turn, leads to everything from domestic violence to war. We have responded to this chronic stress and rage by taking drugs and buying products that stimulate these ancient parts of our brains.

Nor are stress and anxiety (and their bestiary of associated disorders) the only manifestation of our ancient urges to flee or freeze in the presence of predators or other dangers. "Phobia," after all, is simply a word for fear that is directed against something one should not fear. Our modern phobias relate to our ancient system of fear, misplaced in our modern circumstances. Along with these phobias come panic and even post-traumatic stress disorder, both also related to fear cues left to wander, misplaced, among the cells of the mind.

Some people are more susceptible than others to the modern consequences of tigers, leopards, and our residual fear. Part of the difference from one person to the next is genetic, another part relates, complexly, to experiences, whether in childhood or later. Maybe you are lucky and you sleep easy, afraid of nothing, no rage in your heart (or rather amygdala) and no fear either. Or maybe you live a life in which your fear still makes sense. But if so, your company is sparse. The response of most of our adrenal systems in the context of our modern fear and aggression is no longer adaptive. It is out of context and control and so without recourse we tend to medicate ourselves, whether with prescriptions (which can help with anxiety but not panic or phobias) or street drugs, to quiet the predators that still pace our brains. The medications cost us billions a year. The street drugs cost us far more—economies, livelihoods, and lives. In the future, it has been suggested, we might be able to silence the actions of the genes that make us fearful, phobic, anxious, or angry, to teach ourselves, in other words, genetically, that the tigers we imagine everywhere are gone. Meanwhile, we have nearly extinguished the real tigers from the real world. At zoos, we press our faces up against their cages to be reminded of what they once inspired in us. We laugh when they bat at their bars. Yet we also get the chills, because deep down, under our fat and muscle, our bodies remember. Our bodies remember even as our conscious minds have forgotten. So it will long be. Long after even more of the tigers have been left to go extinct, we will remember them, one sleepless, adrenaline-primed night at a time. Nor is this discontent even the end of the story. Predators shaped our fears, but that is not the limit of their influence. The legacy of their brutality leaves many marks, the most pervasive of which frame what we hear and smell in the first place, how it is, in other words, that we construct and live in the world. They shaped everything that, through the lenses of our senses, we sought to change.

11

Vermeij's Law of Evolutionary Consequences and How Snakes Made the World

Imagine it all otherwise. Imagine that you were able to see smaller or more distant things than you do. Imagine that you had a better sense of smell. Each species constructs the world out of the signals received from its senses. Birds and bees see ultraviolet patterns. Ants see the stripes of polarized light in the sky. Vipers see heat, taste chemicals in the air, and feel the vibrations of each banging footstep through their skin. We do not experience their perceptions, except through tools we have invented but not internalized. The world our senses have created in our minds is visual, the other senses only secondary, unnamed characters in the big Hollywood plots of our lives. Look at the chair you are sitting in. Look at the walls around you. They were chosen for their visual appeal and, perhaps to a lesser extent, their texture. They were not chosen for their taste or smells, or by any of the visual cues that would be apparent to other species but not to us.

Our eyes do not just guide us. They lead to our actions. Children choose among beach shells by their colors and appearance and we have, collectively, chosen among the living things on Earth in similar ways. Wild roses smell lovely, but the roses we have tamed and bred are essentially unscented. We chose visual beauty over aromatic beauty because of the quality of our eyes. We make such choices again and again. We chase the large, visible animals, particularly those perceived to be dangerous, like coyotes, out of our cities, but pay less attention to the smaller and less visible species that cling to night and walls. We kill innocent rat snakes in the garden because they are big and black and visible, but miss most roaches and bedbugs, to say nothing of even smaller lives, as they sneak past us unnoticed, species that would be obvious if our senses were different. We ignored the microbes until someone told us they were everywhere and then we overreacted toward them (though only the species susceptible to our drugs). In other words, all of the ways we have changed the world, changed in particular the species we interact with, are because of our vision. What is more, while our vision has become more dominant, at least some of our other senses have actually atrophied. The genes for smell have been left to wither and break. We can distinguish fewer smells than could our ancestors. But our eyes are marvelous and powerful. The question then is how such eyes and their influence evolved. Right now, as you read, your eyes, eyes that evolved in Africa under the tropical sun, are distinguishing the fine lines of letters, running up over *r*'s and down and around *u*'s. The abilities of our eyes and their influence deserve explanation. That explanation, to the extent that we have it, seems to have a great deal to do with a woman named Lynne Isbell and snakes.

Lynne Isbell is a primatologist at the University of California, Davis. For most of her life she had no intention of thinking about how her own blue eyes evolved, much less those of the monkeys she studied. Then one day she was moving too fast as she ran through the forest chasing monkeys. We tend to think of ourselves as the successful primate and yet as she struggled to keep up, it was hard not to feel as though she was the one who had fallen. Her body was slow, awkward even, in the grasslands of its origin. She high-stepped over logs and branches and listened for the sound of leaping monkeys. Then it happened. Midstride, she realized she was stepping on a thin black snake across the path but she could do nothing to change the outcome. Adrenaline flooded her body. Fortunately, the snake, perhaps a cobra, did not strike. It went on its way, a little worse for the footwear. This would not be her last close encounter. In the coming years, she would come face to scaled-face with a cobra in defensive posture and later a puff adder. What amazed her in these cases was that she saw these snakes at all and then that, in most of the cases, her body, somehow, responded to seeing them before she was aware of them consciously. It was as though while her intellectual self had been looking far ahead, another more self-conscious self had been looking all around. Lynne froze just outside of stepping-on-snake distance before she even knew why she had stopped. That she detected these snakes and responded before she became aware of them was a kind of mystery of vision, the brain, and fate. Her experiences were not quite life or death, but that mystery would eventually change her life.

Before the snake encounters, and even for a few years after them, what Isbell had planned to do with her career was to carve off a quiet area of intellectual space and spend the next couple of decades studying within it. She was interested in the social behavior of monkeys, including their peregrinations (hence, in part, her pursuit). She wanted to understand why the female monkeys of the Americas—spiders, woollys, squirrels, and all the rest—often move away from home upon reaching adulthood, whereas those of the Old World (Africa and Asia) almost never do. This was not the only curious difference between Old and New World monkeys. Old World monkeys never evolved prehensile tails. They also have color vision that spans the same spectrum as our own from red through orange, yellow, green, blue, indigo, and violet. Many of the New World monkeys of the Americas, on the other hand, have long, grasping tails and, for most species, an inability to see reds and oranges. These are interesting differences, but in the beginning Isbell focused on dispersal of young monkeys away from their mothers. In the broader story of the history and particulars of life on Earth, monkey dispersal is a narrow topic, but it was enough to intrigue her. Then, through her own primate eyes, she literally and figuratively saw the snake. It was a single observation and yet it primed her the way that, when given gas, a pilot light can feed a much larger flame.

What struck Isbell was a strange little paper about an even stranger disease. She had been trying to understand the evolutionary history of predators and primates, and so was reaching through publications that related in any possible way to the topic. Libraries are full of such bits and pieces of understanding left there for someone to assemble into a story, to put back together in a way that makes sense. The authors of the paper argued that carnivores and monkeys share a specific kind of virus called an RNA retrovirus (HIV, for context, is also an RNA retrovirus).[1] That cats and monkeys would have the same virus suggested one of three possibilities: (1) somebody screwed up in a lab somewhere; (2) a cat had eaten a monkey and in the process contracted the virus; or (3) monkeys have more unusual sexual proclivities than have been acknowledged.

To Isbell, a cat eating a monkey was the most likely of these scenarios. It was the only scenario that occurred to her at the time. Isbell herself had lost whole groups of primates—sometimes primates she had named—to leopards. She had also written an important paper on the influence of predators on primate behavior and evolution.[2] All the shared virus really suggested was that something we know happens now had been happening for a long while. The virus was a vestige of a long ago interaction, a living fossil of the sort found in many animals. Isbell paused with the paper for a moment. She held it up and turned it over the way Indiana Jones might look at a clue to a treasure. Maybe it meant something more. She did not quite know what, not yet.

That paper led Isbell to another paper, this one even more peculiar. It noted that one kind of RNA virus found in Asian monkeys had its closest relatives in a snake, Russell's viper (*Daboia russellii*).[3] Russell's viper has, by some counts, killed more humans in recent history than any other snake. It is both winsome and irritable, as its kind may long have been. The authors of the paper had not bothered to discuss why or how this had happened. Had a venomous snake bitten a monkey and led to the passing of the virus a long time ago? She could not prove it. Yet together these two papers suggested to her a long history of the evolutionary game of venomous snake, cat, and poor besieged monkey. Venomous snake bites monkey. Cat eats monkey. Monkey eats fruits and nuts. The passing of viruses from primate to predator confirmed for Isbell the story of monkey as prey. It was not yet her bold new theory, but these were the first pieces, pieces that were beginning to seem as though they had something to do with her own experiences with snakes. She had been on that trail to learn about monkeys but she began to wonder whether what she had found was a story about humans and herself.

Still focused on monkey dispersal, Isbell continued. She decided to study more about snakes and their history and geography, particularly as it related to primates. She called Harry Greene, a Cornell professor and the snake biologist's snake biologist and asked him about the history of snakes.[4] She wondered, as she talked to Greene and read more, whether snakes might somehow explain some of the mysteries of why primates are different in different regions. "What if," she thought aloud to her husband, "the fact that female monkeys are more likely to leave home in the New World has something to do with the density and history of venomous snakes in the New World?" If monkeys were more likely to meet snakes in the Old World, they might also be less likely to recklessly move long distances. Suddenly, her day job and her wild ideas had begun to intersect. It was a point of excitement. She thought about the idea in the car. She thought about it while walking into her office. She thought about it when talking to students or when sitting at the dinner table with her husband. Like a drug, Isbell's innocent theory became irresistible.

What Isbell began to wonder was whether venomous snakes influenced the trajectory of primate evolution, not through temptation but rather through death. She wondered if some of the differences between the monkeys of the New and Old Worlds and around the world more generally were due to differences in the likelihood of being killed by a venomous snake—the persistent effects of the idiosyncratic distribution of life.[5] What if not only the sedentary lives of Old World monkeys but also their better vision and even their relatively greater intellect were all responses to the threat posed by snakes? Perhaps the traits one uniquely finds in Old World monkeys and apes were those possessed by individuals, like her, who avoided the waiting cobra, viper, or mamba. Maybe our better vision had evolved to detect snakes in much the way that our immune system's primary role is to detect pathogens. Perhaps it was to this history that she herself, like all of us the descendant of an African primate, owed her own ability to detect the cobra and other snakes near her. Maybe, just maybe. If so, that was scarcely the end of the implications.

If Isbell was right, and venomous snakes had caused richer vision to evolve in some primates but not others, then she could make a prediction. She had one piece of the puzzle. She knew that in the New World only some primates see the full range of colors visible to the humans, but in the Old World all species do. Could the ancient presence of venomous snakes in the Old World be responsible for these differences? She also knew that in Madagascar, the lemurs, a kind of primitive primate long separate from other primate lineages, not only have poorer color vision but also cannot see in fine detail in the way that we and other Old World monkeys and apes can. Isbell's theory predicted that there are no venomous snakes in Madagascar.

Among primatologists, Isbell's idea has no precedent. But often ideas new in one field are accepted truth in another. One field's radical possibility can be another's dogma. Predation is not the exclusive fate of primates. Being eaten is a very common way to die, whether one is a primate or a mollusk, especially, I suppose, if one is a mollusk. Perhaps the best precedent for what Isbell was beginning to argue came from just such mollusks. It was work done

in the laboratory of Geerat Vermeij. Vermeij works a few buildings over from Isbell in the geology department at the University
of California, Davis, and lives in a house a block away from Isbell's
house in town. They are neighbors in work, in life, and, it turns
out, in ideas.

On any given Sunday you might find Geerat Vermeij at the
beach, on his hands and knees picking through shells. He moves
slowly, like some primitive bird, picking over the detritus, trying
to find the rare or interesting. Vermeij has spent a life among the
shells, whether living or dead, recent or fossil. More than anything,
he has specialized in understanding the diverse ways in which animals die. He studies these deaths as though at a crime scene. Instead of blood and bones, he tends to search for holes in shells and
the suture lines of old wounds. Instead of weapons, he looks for
beaks, raspers, teeth, and the other murderous contrivances of evolution. In this approach, it would be fair to say that he has an uncommon vision of life, except that Vermeij has no vision at all. A
case of glaucoma at the age of three rendered him blind. Doctors
removed his eyes and left him to find his way around the world
with his other senses. Like a snake, he listens, smells, and tastes. It
is his sense of touch though from which he has constructed his most
detailed understanding of the sea and its history. From the shore, he
reaches to the ocean bottom and from there back in time.

One of the mysteries that Vermeij noticed very early in his career was similar to the one upon which Isbell had stumbled, specifically that when new predators—be they crabs, snakes, or modern
man—emerge, their prey change. Since his childhood, Vermeij
has focused on mollusks (clams and snails). In the shells of those
animals, his fingers detect textures, shapes, and nuances of form.
Pause for a moment and imagine doing this work as he does it.
Walk to the cabinets he opens every day. Move by memory among
the echoes of objects. Pull out a drawer filled with shells. Now run
your own digits over them. As you do, pay attention to shape and
size, but also to the bumps, ridges, and twisted intricacies. Notice

too what is missing, though that is the hard part, because to do so, one must know what is supposed to be there in the first place. Notice the gaps, the occasional and seemingly inexplicable hole or divot. Now focus on that divot. Your fingers evolved to pick up fruits and later to grasp stones and spears, but push them to their limits. What could it be? It is perfectly round, this particular hole, as though drilled. But feel inside, push the tip of your pinky down and you might note a rougher texture. All of these subtleties are clues to the history of the shell you hold and, for Vermeij, to the particular stories of the hundreds of thousands, perhaps millions, of shells he has lifted and fondled. Out of those feelings, Vermeij has constructed a world different from the one most of us experience, though no less rich. In his world, some things are obvious that with vision would otherwise be missed.

Vermeij noticed many things in his hours among the shells. He noticed what you might see if you took up his occupation—that shells varied from place to place and time to time. He also discovered other things that everyone had missed. Perhaps some features are more obvious to fingers than to eyes. There were patterns in the kinds of shells one found in different places and times. He enjoyed the differences, the way anyone might delight in finding something new, but also wondered about their cause. It is out of specifics that biologists build their universals, and he was beginning to find himself with plenty of specifics. What was most conspicuous to him was that the species of the Pacific had thicker shells than those of the Atlantic, with smaller, more obstructed entrances and longer spines.[6] What if, he wondered, those differences were the result of differences in the mollusk-eating predators present in the different oceans? In the Pacific, crabs have bigger claws that are better able to crush an unprotected snail. Yet as he felt his way, there was an even more pronounced pattern at his fingertips, a change not across space but through time. Even as the dinosaurs were going extinct, bigger revolutions were occurring underwater. These seemed, to Vermeij, to have at their root not meteors or some other cataclysm, but instead, once again, the specifics of predators. After crabs and other predators arose, the life on the seafloor responded. It had to. Shells thickened. The openings of shells became narrower and everything became more spiny, armored against fate.[7] Mollusks moved from the seafloor down into the sediment. Whole lineages of life disappeared. But there was more. Unlike the fate of dinosaurs, the deaths of mollusks left evidence of the crime, cracks in shells, holes, and other telltale marks that suggested not the two hands of some god, but instead the millions of claws of crabs. In handling the shells, what Vermeij realized was that the evolution of the killing tools of predators had shaped the entire floor of the sea and its inhabitants. Just how the mollusks, his focus, changed was a function of the particulars of what the claws of a given region and time could or could not break open. All of this came together to suggest to Vermeij a kind of law, a rule about life as general as the physical laws are about particles, a rule that applied to mollusks, but also to everything else including snakes, primates, and you.

Vermeij's law was about a kind of natural gravity, the force predators exert on prey.[8] When new predators evolve or when old predators grow more abundant, prey respond. They must, just as clouds must part as they blow over buildings and wet clay must give to the pressure of hands. What Vermeij had uniquely noticed was that the ways in which they responded were predictable and inevitable. To most people before Vermeij (to the extent that anyone had considered the question), it had seemed that prey should respond to those things predators are best at, their most predictable and deadly tools. Vermeij thought the opposite. Imagine a crab that proceeds through four acts in its performance—finding a mollusk, picking it up, breaking in, and then actually killing and eating it. At some tasks, it rarely fails. It finds prey without trouble. It kills with deadly certainty once it has broken through the shell. What it most often fails to do is to get through the shell. Breaking in is hard to do, and so what mollusks have done over time is to change most in those features that prevent the crabs from breaking in. This was Vermeij's law: prey respond to predators' weaknesses, the ways they fail rather than the ways they succeed. The main caveat is that the prey must vary genetically in traits related to the predators' weakness, but in most cases they do. Now that crabs are everywhere, almost all shells in the ocean are thick and hard, but the fleshy bodies of the animals inside are as undefended as the soft bodies of human infants. Crabs rarely fail once they get inside a shell, and so most mollusks do not even try to offer a defense.

Lynne Isbell was beginning to figure out just how snakes fail. It was very different from the ways in which the other predators of primates do so. When lions, leopards, or tigers fail in their attacks on primates, it tends to be in the ambush. These carnivores consistently find primates by scent (we primates are a stinky lot), but they require surprise for a successful attack. When monkeys detect a leopard, the leopard may even turn away, much as the tiger turned away from Corbett when he turned to look at it, as if to simply admit that the jig is up. Without the element of surprise, a big cat is much less able to kill its prey (though sometimes they will try anyway, hunger being hunger). As Vermeij himself has pointed out, large cats fail to kill their prey as much as half of the time once they have detected it, in large part owing to losing the element of surprise. Because of this, monkey responses to leopards have evolved in such a way as to alert carnivores that they have been detected. Many primate species, including Diana monkeys and Campbell's monkeys, scream out alarm calls for "large cat." When doing so, monkeys appear to signal not only to other monkeys but also to the cat itself. So useful is this notice of an ambush that several monkey species are even able to recognize the "large cat" calls of other monkey species and in hearing them know what to do, which is to first look down.[9] Alarm calls are at the core of primates' ability to escape carnivores. They have even been argued to be the precursors to language in humans. My daughter's first word was "fish" (perhaps she was imagining a very big fish), but our lineage's first word may very well have been "leopard."

Chimpanzees also hunt monkeys. Whatever squeamishness you might feel at eating an animal that looks back at you with eyes not unlike those of a human child, chimpanzees do not feel it. Some chimpanzees eat many monkeys but in doing so, they fail at something different than do leopards. Once those chimps detect monkeys, they catch and kill them nearly all of the time—chimps hunt actively and pursue. But they are not very good at detecting monkeys. It does not pay for monkeys to evolve defenses against chimps, and it definitely does not pay to sound the alarm. As a consequence, when monkeys spot chimpanzees they respond by running away or huddling close to branches and becoming totally silent, engaged in a life-and-death version of hide-and-seek.

Then there are the snakes. Snakes eat monkeys, but they also kill them in self-defense (because, after all, every so often monkeys kill snakes). Once monkeys spot snakes, alarm calls are useful to tell other monkeys where the snakes are. Several monkey species produce snake-specific calls—"snake, snake, snake"—and may even do so in a way that distinguishes different kinds of snakes (Campbell's monkeys, for example, sound an alarm when provided with models of Gabon vipers, but not black mambas). There is a difference though between spotting cats and spotting snakes. Monkeys need to see cats, but best at a distance. Spotting snakes at a distance is not necessary. If Vermeij's law is right (and if, as tends to be agreed, many monkeys die from snake bites), then monkeys should have evolved the ability to notice snakes even as the snakes lie motionless and camouflaged. In other words, Old World monkeys and apes should be better than other animals at detecting snakes. Such a possibility was not considered, not, anyway, until Lynne Isbell stumbled upon it, feeling around in the darkness of ideas the way Vermeij feels around.

If Isbell was right that the particulars of primate vision evolved in response to the presence or absence of venomous snakes, she would expect better vision with greater exposure to venomous snakes. That is just what she found. Venomous snakes evolved in the Old World, and were relatively recent arrivals (10–20 million years ago) in the New. This matched the differences in primate vision. It fit her theory. But what about Madagascar, where prosimian-present primates have relatively poor vision? From the beginning, Isbell had hoped, in a way, that she was wrong. If she was wrong, she could get back to the life she was living before her idea. Maybe she would find that there are venomous snakes in Madagascar, but just as she predicted, there are not. Madagascar has no venomous snakes, and Madagascar's primates, the lemurs, have the worst vision of all the primates. They are as likely to find their way by taste, smell, or touch as by sight. In this, they are like Vermeij.

Isbell has elaborated her theory in detail in her book *The Fruit, the Tree, and the Serpent*, and at least two things have emerged as undeniable.[10] First, our color vision, and the color vision of African and Asian monkeys and apes more generally, deserves explanation. The only other prominent explanation for patterns in color vision, aside from Isbell's, is that our color vision evolved to help us discern different kinds of fruit.[11] This seems possible, though it is unclear why color vision would be important for fruit in the Old World but not in the New and not at all in Madagascar, where most lemurs eat fruit. Yet, even if the fruit hypothesis were right, we are still left with the certainty that we have good color vision because of our interactions with other species. Second, once the full spectrum of our color vision evolved and our other senses faded, many consequences ensued, both for us and for the rest of the living world.

In concert with the development of our vision, our brains began to expand. That visual and language abilities, both plausibly linked to our evolutionary relationship with snakes, were at the core of this early expansion is beyond doubt. Trichromatic-color vision and antipredator calls seem likely to have been necessary first steps along the brain evolution trajectory that would eventually make us smart enough to be able to type "brain evolution trajectory." In fact, our vision would become the dominant sense linked to our burgeoning brain. Evidence from the genes of different mammals suggests that just as our vision was becoming ever better, at least some of our other senses grew worse. Genes associated with smell mutated, one after the other, and because smell had become so unimportant relative to sight, the individuals with mutations did no worse. Through time, more and more of our genes for smelling have become broken, unused, and apparently unnecessary, just as has repeatedly happened with vision in cave fish. Whether the same is true for our senses of touch and hearing is unknown, though it seems possible. In other words, for Isbell, snakes are the pea under the pillow of our minds and the ways in which we perceive and have constructed the world.

It is easy to be skeptical about Isbell's idea, just as it is easy to be skeptical about many grand theories in primate evolution. The facts are fragmentary, as they long will be, and the ability to experimentally test theories limited, and so the hands of the archetypal anthropologist start waving as though attached by a string to clouds in a storm. Personally, I wondered about the central pillar on which her entire idea rests, that venomous snakes have killed primates often enough so as to have affected primate evolution. Most snakes on Earth, after all, never kill anyone save rodents and insects. They are shy and reluctant to bite, neither tempting nor terrible.

Yet as Isbell points out, there are many records of primates being killed and sometimes eaten by some kinds of snakes. I decided to do my own kind of test—compelled by something stronger than curiosity. I sent an e-mail to friends asking how many of them knew of a biologist who had made a mistake and accidentally grabbed or had been bitten by a venomous snake. I imagined I would get a list of famous (and famously dead) snake biologists who had made one too many a lunge in the field. Instead, I discovered that a striking number of my friends have themselves been bitten by venomous snakes.

Greg Crutsinger, now a faculty member at the University of British Columbia, was working at La Selva Biological Station in Costa Rica when he stepped over a stick that bit him. The stick was a hognose viper. Greg is still a little twitchy around sticks. Piotr Naskrecki was walking down a trail picking up this and that, looking for katydids and more generally new species. Piotr lifted up a rock and a venomous snake bit him. Piotr lived to discover more species. My former adviser Rob Colwell was walking down a trail and talking when he missed a terciopelo (velvet skin) that did not miss him. It emptied all of its venom into his shoulder, as snakes are wont to do only when they are trying to prey on something or, in this case, someone. Maura Maple, whom I met at a field station in Costa Rica, was bitten by a terciopelo at La Selva Biological Station, not far from where Greg was bitten by the hognose viper. The list goes on. Hal Heatwole, whose office is several doors down from mine, was bitten by a sea snake and took the time to take a picture of the bite because he knew that a friend needed such a photo for a book on deadly bites. Vlastimil Zak, an ecologist who lives in Ecuador, has been bitten at least twice by venomous snakes. These friends lived, but not everyone does. Joe Slowinski, a friend of a friend, went to Myanmar with a team to find new snakes. He is one of a wave of biologists who have traveled to faraway places in recent years to find new things. Someone on the trip, a guide, brought him a plastic bag filled with a snake. Someone thought it was venomous. Slowinski thought that someone was wrong. His taxonomy failed him and he died.

Of course, I know many more people who have been stricken by cancer or car accidents than I do who have been bitten by venomous snakes. Yet in all these stories dwells a basic reality. When biologists muck around in the tropics and fail to pay sufficient attention (or their vision does not allow them to pay sufficient attention), they stand a fair chance of being bitten by venomous snakes (less of a chance than being hit by a car, but cars were no threat during our early evolution). What is more, biologists tend to interact with other species more like the ways our ancestors did—hands on—than do most people. To look at the longer history of human relationships with snakes, it is clear that snakebites were once even more common than they are today, and today they are far from rare. They tend to be underreported, but even the cases that are now number 30,000 to 40,000 fatalities a year, not to mention the bite survivors. A study of more than 1,000 rubber tappers in Brazil found one in ten had been bitten by a venomous snake. Half of those who had been bitten had been bitten twice![12] A seven-year study in Benin tallied more than 30,000 bites by venomous snakes, 15 percent of which resulted in death. An older study from Niger estimated that 10,000 people a year are bitten by venomous snakes in that country. There is no reason to believe these studies are unusual. Instead, for the tropical landscape in which the most aggressive venomous snakes dwell (or dwelled), they seem likely to represent our general susceptibility to dying on the end of a snake fang, particularly when in the African tropics of our origins.

Do I think that snakebites are or were common enough to favor individuals with better vision, vision for seeing still and camouflaged objects? Maybe, especially when one considers the small size of many of our early ancestors and for that matter, their children. As Isbell points out, one of the earliest of our primate kin, *Eosimias*, weighed about a quarter pound, small enough to put between two buns with some lettuce. For such bite-sized ancestors, deaths owing to snakes would have been common, all too easy a way to go. If there was anything different about the genes and traits of those individuals who survived, it would be favored generation after generation. It seems very plausible that what would be different is the quality of the vision and ultimately the brains of those who lived.

In short, I would go so far as to say that Isbell's reading of the history of primates and their vision is both wild and plausible. And, in any case, for my own broader thesis about the consequences of our interactions, it almost doesn't matter. Any answer almost inevitably has to invoke interactions with other species, whether they are snakes, fruits, or something else. I vote for the snakes. Close your eyes and imagine Geerat Vermeij out in the jungle walking down a trail. Beside him is someone with vision. Who do you think is more likely to die of snakebite? Almost inevitably, Vermeij, who can feel trees and know their type, can smell fruits and listen for leopards, but is unable to notice snakes. Unless he can grab them, which is ill-advised, they are not part of his perceptual world. Vermeij has almost been killed by a variety of dangerous animals, for just this sort of reason. He once grabbed a venomous fish, only to have the reality slowly dawn on him as he fumbled with its texture. He also, when reaching for shells, once put his hand onto a stingray's tail. Vermeij, like the lemurs in Madagascar, was fortunate enough to be born where death by snakes is rare enough that he can thrive by feeling his way. Across most of the long history of predators and poor besieged monkeys, he would not have been so lucky.

Once we had better eyes, the aspects of the world that were apparent to us changed, and as they did, the world itself changed. Our vision may have been shaped when we were prey, but its greatest effects came once we had turned into predators. Its effects included nearly everything, good or bad, that we would have done to the world, including the changes we have instigated in our relationships with predators and snakes themselves. When crabs and their claws evolved, the rest of the sea's creatures changed in response. When Eve looked at the snake, it tempted her to the apple and then, ultimately, to both knowledge and choices about the fate of others. When we evolved to be able to see the snake, we too found the apple, or at least the road to consciousness, tools, power, and consequences. When combined with our preferences, our senses are what frame many of the decisions we make in the world. Even these preferences evolved to help us survive in a world of species that mean to do us harm. They rise up out of the deep well of history and help us choose among the things we perceive. Our preferences are not usually conscious and yet determine much of how we act. We and our actions are tethered to who we were, pulled this way and that among the scenes our eyes relay, colorful scenes no longer filled with snakes and yet still shaped by them, regardless of who we are, every day. It is our senses and preferences that would, in the end, lead us to begin killing snakes. We kill them regardless of their danger because we see them. They suffer from what they have done to us, letting us see, however unclearly. In some places we learned to distinguish the deadly from the innocuous or, more simply, to avoid snakebites (the invention of rubber boots saved many lives). But in other places we thrash blindly with our shovels and machetes, and so the snakes suffer the consequences of who we were. We gave not into temptation, but instead into our senses, led by our eyes this way and that through the apparent world.

12

Choosing Who Lives

A handful of scientists spend their days trying to identify things universal to all human cultures that have ever existed. They paw through ethnographies and search anthropological studies for the differences between one tribe and another, but also for those things that are not different. They look for the ways that they themselves are similar to people in Tahiti or Timbuktu. These scientists have compiled a list of a few hundred or so attributes of humans that apply to very nearly all of us, whether we live in a tree house in Papua New Guinea or in an apartment overlooking Central Park. These similarities are at the core of what unites us, despite our differences. Among these universals is a tendency toward wariness around snakes. But there are others too. We all seem to be fond of sweets, salt, and fat, and have an aversion, at least at birth, to bitter foods. Even more intriguingly, most humans appear to prefer scenes with a bushy tree out on an open plain with a little water somewhere in the distance. Most, and perhaps all of these universal preferences relate to our evolutionary past where such fondnesses evolved and made sense, even if they no longer do. These universals follow from the ways in which our senses affect our perception. Universals would be quaint if they had not shaped the version of the living world we have constructed around ourselves. They would be quaint, that is, if they were not the source of many of our real problems, especially those associated with how we have changed the world.

As we walk past each other on the street or look through each other's car windows in passing, we tend to assume that other people are like us. The very similarity of what we see others doing, relative to what we do—walking, driving, spitting, and grimacing—suggests our deep sameness as does the sensation we derive from reading a familiar poem. In the ancient cave paintings of Namibia, one finds hunters chasing prey. These hunters have bodies like our bodies. We cannot help but feel that the person who created this, and his or her intended audience, was very much like us. We feel connected in our essential humanness. Yet the truth is that most attributes of our thoughts and behaviors that one can enumerate differ from person to person, or culture to culture. Some of us have gods. Others do not. Some of us have a single wife or husband. Others have many. In some places, violent reactions to affronts to one's pride are the norm. In others, pride scarcely exists. In some places, fat ankles are beautiful, in other places fat behinds. In relatively few cultures, thinness is sexy. We are all the same species and yet, because of the speed with which culture changes, the variety of our settings, and the whimsy and idiosyncrasy of history, we do and like different things. In fact, the most remarkable thing about universals is that they exist at all, so much do cultures vary from one place to the next. Relatively few of the truths we hold to be self-evident are held to be so everywhere.

Given that there are nearly 7 billion of us on Earth with an infinite capacity for variety, that few of our attributes are truly universal is interesting. These attributes must ultimately follow from our biology. If any of our universals were within our conscious control or could drift with the shifts among cultures, they would have varied by now from place to place, even if only because of rebelling teens. That a few have withstood the chaotic consequences of our variety suggests they are both genetic and, in some deep and primitive way, beyond our ability to change.

Our visual preferences, those shaped by the influence of preda-tors and snakes, are the most pervasive in their influence. But they are also complicated. Perhaps the most straightforward of our evo-lutionary preferences is taste. If we understand taste, we can use it as a kind of model for beginning to understand vision. Stick out your tongue and touch your finger to it. You will feel two things at once: Your finger will feel your tongue. In turn, your tongue will feel and taste your finger. Just which tastes it picks up depends on your finger. Five basic possibilities—sweet, salty, bitter, umami (savory), and sour—can come together to yield more nuanced im-pressions (index finger with a hint of peanut butter, perhaps?). The taste buds themselves look like brain corals at the center of which are the sensitive tips of taste cells, each of which ends in a thin hair. When you eat, little particles of food are washed over these hairs. If a sugar washes over a sweet taste bud's hairs, a signal is sent by the nerves under the taste bud to your brain. The hairs are stimulated and the chemical chain is yanked until it rings somewhere in the space between your ears, "sweet." At least two kinds of signals are sent when "sweetness" is sensed. One signal is sent to your con-scious brain that triggers the sensation you think of as sweetness. Separately, a signal is sent subliminally to your older, deeper reptil-ian brain, where it stimulates hormonal changes in your body in response to having received sugar.

With its taste buds, the tongue is a gourmand's muscle. We have grown used to our tongues and what they do. We take them for granted, leaving them to wallow for whole afternoons in coffee or bad wine. To me, though, what is most interesting about our tongue is that it influences us, in the way that a political tail can sometimes wag the dog. After all, the sensation of taste is a trick. The categories of chemicals our tongues distinguish (sweet, sour, etc.) and how they "feel" to us are produced in our mind. Cats, be they domestic or wild, have a nonfunctional gene for the receptors associated with sweet food and so they never experience the sensa-tion of sweetness. We might have evolved to detect different groups of compounds with our tongues. Or our perception of those compounds that we do detect might have been very different. Nothing about sweet foods gives them an inherently sweet taste. The entire concept of sweet and its taste is an evolved product of our mind. But why? Why have some of our taste buds evolved to signal tastes that our minds experience as pleasurable (sweet, umami, and salty) while others signal something more ambiguous (sour) or even downright bad (bitter). Why do our taste buds produce the sensation of taste at all?

One could imagine, as a thought experiment, scenarios in which our taste buds might produce no conscious sensations. If their only purpose were to modulate our internal hormones and digestive enzymes our taste buds would have no reason to notify our conscious brain that something had been tasted at all. This is just what happens in our guts. As late as 2005, no one knew we had taste buds in our guts. It now appears that that is where most of our taste buds, or at least taste receptors, reside.[1] Those receptors are identical to the ones in our mouths with two exceptions—they are arranged in smaller and more diffuse clumps, and they are not wired to our conscious brain. As a result, they send all of their signals to the subconscious part of our nervous systems and bodies more generally. When food makes contact with them, the taste receptors in our guts initiate waves of response around our bodies. They can trigger salivation, but also other diverse consequences.

Although they are triggered subconsciously, the effects of the taste receptors in our guts are visible to us. We can see them at work when we eat noxious foods that make us vomit. The reflexive opening of our mouths and expulsion of food can be triggered by the response of the bitter taste buds in the stomach that respond to tastes they interpret as toxic. None of this surfaces on our conscious minds until we find ourselves facedown in a toilet. The taste receptors in our guts are evidence of the ways in which all of our taste receptors and buds might have worked. That the taste buds in our mouths lead us to become pleased or displeased is because of our ancestors. Those ancestors whose taste buds triggered a pleasant sensation when they ate foods they needed to find more of were more likely to survive. The reverse holds for dangerous foods and unpleasant sensations. Like lab animals, our ancestors were trained by their sensations to chase after some things and flee others. Their tongues praised them into the right decisions: "Look for more of this sweet and you will be rewarded!" But they also scolded them out of wrong ones: "Put that plant in your mouth again and I'll make you suffer. I swear to god I may even make you puke."

The reason, then, that taste buds elicit sensations in our conscious minds is to trigger preferences and ultimately actions. It is for this reason that our taste buds are so weighted to just a few good (sweet, savory); bad (bitter, sour); or slightly more complicated (salty) sensations. We all prefer sweet and savory foods because we all have those same taste buds. We all enjoy salt until it is too concentrated, for the same reason. Taste buds produce innate preferences because they evolved to help us to distinguish things that we needed from those we must avoid. Bitter and sour have long triggered aversion (whether in fruit flies or humans) while sweet, savory, and, for the most part, salty trigger us to find and eat more.

The problem with our taste buds—and, I will argue, the problem with most of our universal preferences—is that they evolved to favor things that were rare and necessary and to disfavor things that were bad, in a context far different from the one we now face. This system of goods and bads—a kind of sensory morality—worked for hundreds of millions of years. It is analogous to the systems bacteria use to move away from bad things and toward good ones. Like the bacteria, we moved toward sweet fruits, fatty meat, and salt (be it in dirt or anywhere else) and away from what was deadly or poisonous. What changed was that we invented tools and gained power over entire landscapes. We developed the ability to make common what was once rare. But it was not simply that we farmed. Other species farm, be they ants, bark beetles, or termites. We combined farming with the ability to process foods, to extract specific compounds and flavors and so to feed our taste buds in ways that stimulated the taste buds, without also bearing the nutrition such tastes once signaled. Consequences we were not smart enough to anticipate ensued, consequences like those that would befall an African bird, the honeyguide. Its small canary-sized body is a measure of a much broader problem—that of sweetness, desire, and the fate of the world.

The honeyguide lives in much of Africa, where it eats the wax, brood, and eggs of honeybees. In this, it is relatively unique. Wax is indigestible to most animals. The honeyguide has been simultaneously blessed with the ability to eat wax and cursed with the dilemma of how to obtain it. Honeyguide beaks are too small to break into beehives. Humans have a different problem. We crave beehives for their honey. We are willing to do almost anything to get to honey. In Thailand, little boys are sent a hundred feet up into trees with a smoking stick to do battle with three-inch-long giant bees and take from them their honey. All over the world children, men, and women have found themselves face-to-face with bees, deep in hives, covered in stings and yet, pleased with their sticky, sweet discovery. Honey, to paraphrase the anthropologist Claude Lévi-Strauss has "a richness and subtlety difficult to describe to those who have never tasted [it], and indeed can seem almost unbearably exquisite in flavour . . . [It] breaks down the boundaries of sensibility, and blurs its registers, so much so that the eater of honey wonders whether he is savoring a delicacy or burning with the fire of love." The problem for humans, though, other than the stings (which we have learned to more or less avoid) is in finding hives. Together, honeyguides could find hives and humans could break them open, which could yield a sweeter life for both man and bird. So it was that over hundreds if not thousands or even hundreds of thousands of years, the honeyguide and East Africans came to realize each other's talents and to depend on one another, bird and human.

Many bird biologists have watched the interaction between the greater honeyguide (*Indicator indicator,* the name itself an indication of its story) and humans. A honeyguide, when it has found a hive, will come to the nearest house or person. There it will call, "tiya, tiya," flash the white of its tail, and fly toward whoever is lucky enough to look on. It will continue to do so until someone follows it to a hive. At the hive, it will call again and wait. With luck, the hive is low enough to be climbed to, whereupon the person, a gatherer of honey, finds a food that rewards his or her sweet taste buds and the honeyguide finds a taste that rewards its too (our taste buds are sufficiently ancient that we and the honeyguide have similar fondnesses).[2] No other mammals are known to follow the honeyguide, and so every bit of its elaborate act seems to have evolved for us, that we might help it and it us to sate our respective taste buds. Then, very recently, everything began to change.

Thousands of miles away from the honeyguide in the year AD 350 (give or take), Indians figured out how to farm sugar in the form of sugarcane. Over time, the process grew in sophistication until pure sugar-crystal sweetness could be extracted from the cane. It was, in the history of humans, a revolutionary process. What was once valuable because of its rarity became common as sugar cane and the ability to process it spread. Elsewhere, sugar beets were domesticated. Each year we have farmed more sugarcane and sugar beets. Now they are joined by corn farms. On such farms, a useful food (corn) is farmed to produce nutritionally useless sweet high-fructose corn syrup. In 2010, more than 400,000 square kilometers of Earth were dedicated to the farming of sugar beets and sugarcane,[3] an area the size of California. A similar quantity of land is dedicated to the corn used to produce corn syrup.

When millions of humans continue to starve each year, the fact that we have allotted an area this large to a substance for which none of us has any real need (even without adding sugar, all of our diets now certainly have enough of the stuff) is a sign of just how beholden we are to our taste buds. We might see our investment in sugar agriculture as a choice, but it is just as reasonable to see it as an inevitable consequence of what our taste buds can perceive and what they tell us is "good." Because we never, in our long evolutionary history, faced a situation in which we had too much sugar, we have no bell or whistle in our body that tells us that we have eaten too much. Our body's demand for sugar is essentially infinite and irrational, but that was never a problem until we evolved the ability to wield tools and change the land.

Back in East Africa, no one follows the honeyguide anymore. It has stopped coming to villages. The children who once chased it pursue lollypops instead. We traded partners, so that as much as honeyguides are now more rare than they used to be, sugar beets and sugarcane stems are common, more common even than humans—thousands of stems for every woman and man on Earth. No one chose to ignore the honeyguide; we simply did whatever was necessary to keep our taste buds happy. Because of this tyranny of our taste buds, those few species that could provide us with lots of sugar were favored, and those that provided us with a beehive here and there became more rare.

In many places in Africa, even where no one collects honey anymore, there are still stories about the honeyguide. They say that if anyone collecting sweets ever failed to reward the honeyguide with a little sugar and wax, the honeyguide would turn on them. It would take the elephant, or maybe the hippo, to beehives, and abandon humans altogether. No elephants have been seen following honeyguides, but moral stories can speak to the nature of a consequence even if the specifics are wrong. We have failed to reward the honeyguide, and we have borne the consequences, albeit from too much rather than too little of all that is sweet.

Just as we once needed sugar for energy, we long needed salt for reasons of historical contingency. Our circulatory system evolved when we were still fish in the sea, and salt was everywhere. In that context, evolution favored salt and other common compounds for the core switches, levers, pulleys, and other parts of the body. Salt, in particular, was used throughout our bodies. It helped them to regulate blood pressure, which is still one of its main functions in our bodies. Other nutrients might have worked, but in the sea, salt was cheap and easy. Then we left the sea and moved ashore, where salt is scarce. We searched it out, as did other species. Macaws fly to saltlicks, elephants walk to them, and pregnant women can sometimes be found eating fistfuls of salty clay. It was during this time of transition, from oceans to land, that our taste buds for salt became refined and accentuated. The cable that links the taste of salt to our pleasure response was wired deep and strong. It was easy to die without salt. Our brains needed to remind us to search for it.

In the last several hundred years, our needs for salt changed, just as was the case for sugar. We developed the ability to harvest, store, and even produce salt. Now we are like fish again. We have plenty of salt, but our taste buds are ancient, and so they still beg for the stuff and we give it to them, one chip, bowl of tomato soup, or soda at a time. Our salt taste buds, unlike our sweet or savory taste buds, have limits. We perceive salt concentrations that are too high as bad, but below that concentration we crave more and more. You might blame yourself for your inability to stop eating salty foods, for a lack of self-control perhaps. But the truth is that you are doing what your body evolved to reward you for. Your salty taste buds have no job other than to remind you of the goodness of salt and that you should find more. They beg it of you. Part of the struggle, then, to control the intake of sugar, salt, fat (which the savory taste buds holler for), or anything else, is that while your conscious brain may tell you to avoid it, elsewhere your brain stimulates you to search it out. Ours is a universal struggle not of will power, but between who we are and who we were.[4]

Nor is it just salt and sweet. Our other taste buds beg, universally, for fats and proteins, also limited for much of our history. As for bitter and sour flavors, they do the reverse. Bitter flavors signal strong chemicals, and so when we taste them, we have a bad sensation, a sensation that makes us want to spit or vomit. Our bitter and sour responses are actually triggered by many different compounds, compounds that share little in common except for being toxic. Our tongue is amazingly sophisticated. It assesses complexity in the world and simplifies it for us into two possibilities: find more or spit it out. We vomit up bad foods because our body would have it no other way.[5] It has saved us, one nasty berry or leaf at a time. So it is that throughout our long history, in a world of toxic and necessary things, we survived.

Our taste buds are a good point of departure into our preferences more generally, because they evolved for no other reason than to lead us to what we need. Our taste buds, like other aspects of our preferences, are very different from hunger or thirst. Taste buds do not tell us how much sugar or fat we need, or when. They evolved without shutoff valves (other than those related to satiation more generally), and work on the evolutionary "assumption" that we always need both substances. No matter how much you have eaten, when you touch a piece of cookie to your tongue, your brain will ring out "sweet." Thirst and hunger are different. They tell us when we need water or food (in response, in part, to sensors that measure how much our stomachs are stretched with food). Once our body has enough, or once it thinks it has enough, it stops asking for more. One can even inflate a balloon in the stomach and it simulates the same effect, the same contraction-like feeling of fullness. Not our taste buds; they tell us what we needed a thousand years ago and we oblige. We could be at the point of dying of high blood pressure—collectively we are—and our taste buds would still say, "Salt is good." They are irrational in their modern context.[6]

But what about those other apparently universal preferences and dislikes that I listed, the ones that relate not to tastes but to sights, sounds, or even smells? Some of our perceptions of odors clearly evolved to point us toward or away from factors that influence our well-being. We universally dislike the smell of excrement.

It seems plausible that the odor we experience when we smell excrement evolved to keep us away from the stuff (it may not seem as though we need a reminder to stay away from piles of poop, but you overestimate your ancestors). To the extent that dung beetles experience an odor in their head when they smell excrement, it is likely a pleasant one, just as for vultures, dead bodies must smell lovely. Dung and carrion do not intrinsically produce bad odors any more than sugar produces a sweet taste. Such are the vagaries of the senses. Our brains respond differently to different sorts of sounds, and although this effect has been poorly studied, one can imagine contexts in which this would have benefited our ancestors' fitness. Much of what we experience as universal seems directly related to our survival, if not now, then once upon a time.

But back to vision. Vision must be different. Whether or not it was shaped by snakes, it is our special sense, our privileged child. Our tongues, noses, and ears respond differently to the various categories of stimuli, but they are marginalized relative to the observed world. Vision responds to intricate scenes in their entirety, whether they are Jackson Pollock paintings or tigers leaping toward us. Our vocabularies to describe touch, taste, and smell are meager. Not so for vision, which we can describe in detail, with shades of color, qualities of light, and hundreds of associated adjectives. It seems implausible that buried in the process of vision are the same kinds of preferences that we find in the way we taste or smell, implausible and yet possible.

We know certain scenes consistently trigger the same responses in individuals, regardless of their cultures. Snakes trigger us to jump back. They trigger nightmares and fear, except where that fear is counteracted early by culture or education. Scenes with water trigger pleasure, as do open landscapes, grasslands, and forests that look as though they have been tamed, with their understories cleared away as if to make a grove. Could such responses also be the consequence of our evolutionary past? Could it be that some classes of images make us all happy and others all scared, and that these effects are, or at least were, adaptive? Once it was useful to fear snakes. Maybe once it was also useful to move toward grasslands and away from forests' dark understories. Here we return to Lynne Isbell's snakes that, like bitter food, once killed us and may have honed our eyes.

It has been difficult for scientists to study the adaptive aspects of vision. We are wooed by our vision, but we also use our vision as the tool with which we study ourselves, and so just as a dog has trouble biting its own tail, we seem to have difficulty seeing our own eyes. Nonetheless, all of the explanations for our sophisticated color vision implicate our ability to detect other living forms, be they fruit or snakes. What if, just as for taste, smell, and maybe even hearing, our visual systems trigger not just our conscious responses, but also subconscious responses to categories of scenes, categories that once saved us and now just influence the world? If our tongue can lead us toward good food, isn't it possible that our eyes might also lead us toward things that will help us and away from those that will do us harm? We seem to simultaneously imagine that our eyes are the most sophisticated sense, and that they are the least able to recognize specific elements of the world's goods and bads, stimuli worth running toward or away from. Maybe our eyes trigger preferences too.

Let's return to the snakes. It is hard, after all, to turn away from them. Our fear of snakes, or at least our initial wariness around them, seems universal. It is easy to believe that when we all lived outdoors in the tropics of Africa and Asia, where death from venomous snakebites or even constriction by snakes like pythons was relatively common, such wariness would have saved lives. Anyone who was universally unwary of snakes (say, for example, herpetologists) would have been less likely to pass on their genes. What is surprising is that this fear persists, regardless of where and how we now live. We are as predisposed to becoming afraid of snakes in Manhattan as in a rain forest in Cameroon. We are more afraid of snakes, on average, than of cars or guns. Not everyone is afraid of snakes, but surveys tend to suggest that more than 90 percent of us are. This fear develops early and is either innate (we are born with the fear) or easily learned. Monkeys who are shown videos of another monkey responding fearfully to a snake will be afraid of snakes for the rest of their lives. Monkeys shown videos with a rabbit in the place of the snake never become afraid of rabbits. Snakes seem to occupy a unique category in primate brains, unlike other threats, and even unlike other threatening creatures. Monkeys do not seem, for example, to so readily learn fear of cats. Nor is it just the monkeys. Infants too young to talk appear to focus on videos of snakes (instead of a video of another animal) innately (which is to say, without learning) when an adult near them is talking in a fearful voice. But when the adult talks normally, the infants give equal attention to snakes and, say, hippos. It is as though our brains have a rule along the lines of "If you don't have anything to fear, don't fear anything, but if you do have something to fear, fear snakes."[7] And so we do.

One explanation for how some scenes trigger negative responses (in the way that bitter foods also do) relates back to the amygdala, that element of the brain that tugs the strings in our bodies when we are chased and when we fight. Hundreds of biologists spend their lives scaring rats to study their fear, amygdalae, and vision. They can tell you that if you show a rat a scary picture (whether it is of a cat or a biologist), it responds, even if its conscious brain is focused elsewhere. Scary images excite rats' amygdalae, but not their frontal lobes (the part of the brain greatly exaggerated in you and other humans and associated with our cleverness). More to the point, when the amygdalae is removed from monkeys, they lose their fear of snakes. For a while, it was not clear whether fear was also being conveyed subconsciously like this in humans.

As they tend to do, brain biologists searched long and hard for individuals, both rat and human, with problems. They wanted to find individuals with "blindsight" who could still see but had no conscious knowledge that they were doing so.[8] Blindsighted individuals are not aware of seeing, just as our guts taste our food without our being aware they are doing so. Individuals with blindsight are typically surprised to realize that they are aware of the locations of objects. In a recent study, a man with blindsight walked down a corridor, zigzagging around objects he did not know were there. (Some individuals also have emotional blindsight and can respond to fearful faces by cringing, even if they are completely unaware that they are seeing anything at all.) That blindsight can occur means that we all experience both conscious and subconscious responses to what we see, just like the rats. This in turn leads to questions about what those subconsciously encountered visions are doing. Further questions arise, such as how and whether blindsight might, like our guts' taste buds, subconsciously register distinct categories of scenes, whether they be specific, such as snakes, or just more general varieties of fear.

Arne Öhman (a brain biologist who happens to be terrified of snakes despite living in snakeless Sweden) and his colleagues have developed a test that mimics the effects of being blindsighted. To do this, they present participants with a picture of a face. Sometimes the face is accompanied by a loud and distracting sound; in other cases, it is not. When the face is accompanied by a loud enough sound and a quick enough image, Öhman can produce situations in which although the image is seen, no image reaches the participant's conscious brain. When asked, the participant says that he or she did not see a face. In addition, the part of the brain that would be expected to light up if a face were seen (in MRIs being done simultaneously) does not. What lights up instead is a separate part of the brain, a part of the brain that suggests that at least some signals pass directly to the amygdala. They are signals no one had noticed until Öhman's work and about which we are not consciously aware, ever.

The chain of connections that researchers have identified are old wires that are more prominent in rats that in humans, but present in all mammals. It is the ancient wiring of fear, aggression, and urges. Some visual stimuli and scenes directly trigger this old wiring, and the body responds to them without the signal's ever becoming conscious. When Öhman shows subjects snakes or scary faces, the signals travel to their amygdala and trigger a general bodily response to fear. This occurs even when the participants' conscious brains were unaware that the participant had seen a snake. They could not reason about the snake, because reason was not even involved.

Just how the old wiring in our brains works is far from well-known, but that it is there, making us jump, shiver, run, or strike something is unambiguous. It does not seem a stretch to implicate this old wiring in some of our preferences (or fears) of images of some animals over others and so too in our views of what is ugly or beautiful, peaceful or terrifying. In rats, there exist cells associated with the old wiring that help to mark how close an individual is to an object. Separate cells, called place cells, keep track of when an animal passes a landmark or the way the face is pointing. How far-fetched is it to think that these cells might also register the signal recorded when the eyes follow the sinewy shape of a snake, that narrow fellow in the grass? How much of a stretch is it to imagine that these subconscious parts of the brain might record even more nuanced aspects of the world, aspects that produce bad or good feelings, anger or joy?

What we know for now is that we all seem to be born with an easy ability to learn or have triggered a wariness of snakes, and an ability, once we become afraid, to develop a heightened wariness, a real fear. Most of us are also born preferring relatively open landscapes over forested ones. A scene with a tree that has branches for climbing is judged by most of us as more beautiful than one that is long and straight. These preferences, like fear of snakes, can be modulated by learning, made stronger or weaker by experience and reason, but they seem to begin deep and innate. There are others too, such as universal preferences for water, or shimmering blue colors. Just how all of this works, which scenes we truly prefer, how they are learned and not learned, and how our body responds to them is fascinating, a primitive gem as fundamental to who we are as our taste buds, an undocumented kingdom just beginning to be explored.

Even as the cogs and gears of our universal preferences remain enigmatic, their consequences are clear. By leading us to our choices, they have shaped the living world and removed us from the nature in which we evolved. It began when we turned from prey to predator, from fear to some more complicated mix of fear and aggression. Just as for the crab, our influences ramified out from our tools and senses. Once we had weapons, the first species that we influenced were those we could see and catch. We sought them because our eyes and ears detected them and because when we did catch them, their fat rewarded our tongues. Survival in the face of our weapons meant and means hiding from our vision or breeding very fast. We chased the big and obvious animals. They may have tried to escape by taking advantage of our weaknesses (as in Vermeij's law), but with spears and society, our weaknesses would prove fewer each year. As we looked out across the landscape, we also began to burn. We lit dry grass, dry leaves, trees, and anything else that would ignite. We turned forests to grass over which we could newly see into the distance. With agriculture, we again chose species that grew in the open and planted grasslands of millet, wheat, and corn. The corn was short, the millet too. Where we could not plant crops, we raised cattle that beat down the tall grass, making even more acres of Earth open and, to most of our sensibilities, beautiful. In some places, we made and make subtle distinctions, for example between truly dangerous and mostly or entirely innocuous snakes. In other places, such as Texas, we killed them without such discernment, as we still do each year in rattlesnake roundups. Each of these changes made the world more like our preferred landscape, the one that brings us pleasure whether we are conscious of it or not, whether it is good for us or the future of life or not.

Grasses and cows were not the only species we favored. We also came to choose species that were beautiful to our senses, be they sweet-sounding birds or brightly colored goldfish. That we see such species as beautiful begs the question of whether beauty itself is like sweetness, an adaptive sensation meant to help us survive. No one knows, not yet. In the meantime, tulips and other flowers are shipped around the world at huge expense. Goldfish live in houses in nearly every country. Dogs, which appeal to our social sense of appeasement and connectedness, were brought into our beds. (Cats—well, no one can explain them.) The trains of modernity grind down the tracks, carrying what we call for, one load at a time.

Beyond the species we have consciously favored (for subconscious reasons) is another group, the group that has perhaps most directly risen to our senses, our pests and guests. They sneak around when we are sleeping or through corners and cracks we ignore. Rats hang close to walls, because it is there that they go unseen. Pigeons and other urban birds nest under ledges, where we will not find them. Nocturnal insects run around our houses. Species such as dust mites and bedbugs, just small enough to avoid being seen, crawl on and off our bodies with impunity. Tiny creatures, bacteria and archaea, flourish. We swipe at them, but they grow on, changed and yet persistent.

Our senses, coupled with our power, changed the world so quickly and universally that it is easy to forget what the world used to be like. Today, roughly 60 percent of the earth's surface is managed by humans for production, and most of that land is devoted to one or another kind of grass. Nearly all humans on Earth live by water. (Think of your favorite beach resort, but then also, for example, Manhattan and Los Angeles.) Many of us tend to live by water because we need it, but also because we tend to prefer it. It pulls at us like gravity and makes us feel good. Once upon a time, though, before modern humans, there were more forests and larger animals. Rats were rare, as were mice and roaches. Even grasslands were not nearly so common, and the flowering plants that have arisen around us had not yet called to our senses. In many places, the coastlines along which we now so easily walk were hidden beyond dunes tens of feet high, dunes that while useful to us in protecting our shores, obscured our views. The views won, and so in general the dunes are gone, reduced to a minor row of hills that does little but allow us to see even more of what our eyes and brains demand.

Not all is left to the fates and taste buds. In small pockets, reason has won out over our urges. We have instituted conservation agencies and plans, public health systems and public sanitation, each of which required us in some way to choose what was reasonable over what was simply appealing. There are real successes, the triumphs not of intuition or our "gut sense" but of the nonintuitive reason that accumulates across peoples and generations. Other times, when we try to figure it out on our own, reason loses out, trapped in a mental world constructed by senses that evolved to detect snakes and fruits, not to puzzle out global crises. Collectively, we have repeatedly tended to make the same decision, regardless of our cultures and differences. We have universal senses and preferences and so we often make universal choices. Aborigines burned down forests in Australia. Amazonians burned Amazonian forests. Others burned in Europe and North America. People burned because they could but also because the result was preferred, everywhere. In some places, this preference for open habitats has been carried to extreme lengths. In the United States, more land is now planted in lawns—perhaps the most perfect manifestation of our minds' preference for simple, unadulterated openness—than in corn. The sugar remains sweet, the salt tempting, and the open scenes of oceans or grass beautiful in a way that cannot quite be expressed.

It all might have been otherwise. It would have been were our senses different, if we were, for example, blind like termites. Termites feel their ways along dark tunnels, smelling and groping through the world, as do moles, mole rats, and other denizens of the underground world of holes and chambers. In their world, light is largely irrelevant. In some subterranean species, the eyes of their ancestors became disconnected from their brains until they disappeared altogether. Once the eyes are gone, the colors once meant for them become unnecessary too. To a blind queen, a flashy king was no more attractive than a drab one. Because all colors are costly, termites have lost their hues. They have turned to ghostly versions of their former selves—the color of an onion's skin. In this alternative world, smells and textures come to rule. Species of beetles, mites, and even fungi sneak into termite nests. They hide there, in plain view. They do not look like termites, but they feel like termites. They smell like them too. As for their foods, the termites use taste and smell to choose those foods that are rotting, which to them must smell some sort of sweet. We are like them in constructing a world catered to our senses, just different in which senses those are.

In the end, we have often been like the termite or the ant, at the mercy of our urging senses. But reason can prevail, so long as we do not "trust our bodies." Our bodies and particularly our senses lie. They are stuck on the front porch rocking back and forth and remembering the old days. And so, when you touch your food to your tongue, savor the pleasure of the good tastes. Culture, of course, can influence how we respond to different tastes, just like different scenes. We can learn to love snakes, just as we have learned to love the stimulus provided by coffee even though it is bitter. Our aversion to snakes and our attraction to sugars, salts, and fats are the murmurings of our history, but such murmurings can be quieted. Our universal fears and ambitions have been our fate, but they need not be. Whatever the right action might be, the honeyguide will not lead us there, nor will our taste buds, which will continue to call out for more, just as our fears cry out for us to fight or flee.

PART VI

The Pathogens That Left Us Hairless and Xenophobic

13

How Lice and Ticks (and Their Pathogens) Made Us Naked and Gave Us Skin Cancer

Removing our worms left our immune systems floundering. Changing our mutualist partners left us with too much of the wrong food, and killing our predators left us with ghosts in our brains and nervous systems—ghosts that make us leap with fear and panic with anxiety. Yet it is the changes in our infectious diseases that are potentially most sweeping in their consequences, which is what brings us to the story of ticks and hair.

Very recently our ancestors were covered in hair (or really fur, hair simply being the word we gave our own fur so as to feel a little special). In the contrast between us and them, our lack of hair is conspicuous. We now recognize the central links between Neanderthals and modern humans. Yet when the Neanderthals in museums are covered in hair, they look nonhuman, more "animal" than cousin.* Our transition, from hairy to smooth bodied and the complicated way we look at our ancestors in the hairy dioramas of our museums, begs the question of just what happened. How is it that we became essentially hairless and that, in the process, many (though not all) cultures came to view hairiness as unattractive?

*This is true despite the fact that the ancestors of most modern Europeans and Asians had sex with Neanderthals; essentially all Europeans and Asians have Neanderthal genes.

Ninety percent of all American women shave themselves in order to appear more "beautiful." Nor is this the extent of our eagerness for glabrous bodies. It is one thing to shave a chin, a leg, or an armpit, but we have come to love hairlessness so much that all over the world, people are having their pubic hair ripped off with hot wax.

It may now seem "normal" that we are hairless. In the context of our ancestry, it is not. We do not actually know whether Neanderthals were hairy. They may not have been, which makes the way in which we stare at them even more interesting. But our ancestors in Africa certainly were, as recently as a million years ago, as were the very first mammals and all of the species in between. Thick hair was part of what made mammals successful. They stayed warm when all else around them had cooled. Early reptiles may have stomped and roared, but when the sun set, so did their body temperatures. Not so mammals, who relied on their fur, along with a more sophisticated heart, to maintain a constant temperature. Fur was an evolutionary breakthrough, a warm hug on a cold day that allowed mammals to live in conditions too hostile then (and now) for any reptiles other than birds.

Our hairlessness has become a source of what we think of as beauty, a reality validated in every *National Enquirer* article about a "wolf boy." It also has widespread consequences for our health and quality of life. It is the reason for the origin of melanin (the compound that, when present, makes dark skin) in sunny regions. The production of melanin in cells just under the surface of the skin evolved in Africa, along with our loss of hair. All of our ancestors produced melanin and so were dark skinned, but when some of our ancestors moved out of hot climates, melanin blocked too much sun. At least a little sun on the skin is necessary for our bodies to produce vitamin D. Dark-skinned individuals in sunless places suffered rickets. They died, and so, with time, pale-skinned genes were favored, not just once but several times independently, with the northward migrations of humans. In other words, the variety in our skin color would not exist were our skin not exposed in the first place by our lack of hair.

But why did we lose our covering? Like so many of our modern dilemmas, it may be because of the species we once interacted with, and how their abundance has waxed and waned through time. Blame it on the ectoparasites—lice, ticks, and flies. In the caves of our origins, they climbed through our hair and bit us, and when they did, they transmitted, into our blood, disease.

Of the 4,500 or so mammal species on Earth today, nearly all of them are furry. Just a small handful of all living mammal species are essentially hairless. I say essentially because even our bodies are not totally naked. You and I are both covered in tiny hairs that bristle impotently when we get the chills, but never keep us warm. Dolphins and whales are smooth skinned. Their sparse hair is related to swimming; hairless sea creatures are more aerodynamic (though this is not the only way to a smooth ride—seals and sea lions are able to achieve the same effect by having very dense fur). The tendency for some marine animals to be relatively hairless suggested to a few biologists in the 1960s that the earliest humans were swimming apes. Perhaps somewhere between monkey and man, we were mermaids. Maybe we began to live along riverbanks and seashores and then started hunting for seafood when other hominids were beating us up and we could not find anything else to eat. We might have eaten shellfish and sea urchins and then, on one long day, taken the plunge all the way into our naked future. One imagines the scene like something out of *The Blue Lagoon*. If we were smooth skinned, perhaps we were able to swim faster and farther to that last urchin, and in doing so, survive.

This theory, though long prominent, does not enjoy much support. It does, however, highlight the extent to which having no hair is an unusual circumstance. Pause and think of other "naked" species, and you come up with the marine mammals and the naked mole rat. Then what else? Very few other species. Rhinos, elephants, and hippos are low on hair, but make up for it by having—like dolphins and whales—thick, insulating skin. Once hair evolved in the first mammals a 120 million years ago, it was very rarely lost.

So, why then did our species become one of the very few mammals to lose most of its hair if it did not involve better swimming? Perhaps hairlessness helped us keep cool and hydrated in the savannas as we ran around on two legs chasing down prey (or fleeing predators). This hypothesis seems plausible, except that evidence suggests that hairlessness may actually make us more rather than less susceptible to dehydration. Also, none of the other primates that moved into open savannas (or into the dry tops of canopy trees) has become less hairy. Nor have other predators, such as cheetahs, that chase their prey by running. Maybe hairlessness is like a peacock's tail or a mandrill's pink butt, useless and extravagant, but lovely and therefore chosen. One can imagine that men have tended to choose less hairy women (or vice versa) because such hairlessness is a demonstration of their good genes, genes so wonderful that their bearer need not worry about sunburn or the discomfort of sitting bare-bottomed on a log. That is what Darwin thought. His own wife had a very smooth face, though one hesitates to draw general lessons about mate preference from a man who married his cousin. But the truth is that one does not have to be terribly fit to be hairless. Instead, the early stages of hairlessness (a little here, a little there) seem likely to have suggested a bad case of mange more than they evinced good health.

My own favorite theory has been independently suggested by three waves of scientists across more than a century. Each of them has argued that hairlessness evolved because our ancestors were unusually tick, louse, fly, and, more generally, parasite-ridden. This idea was first suggested in the 1800s by the jack-of-all-trades Thomas Belt in his book *The Naturalist in Nicaragua*. Belt had done his time in the tropics, where his few hairy parts were invaded by ticks, lice, and other life-forms. His own repeated infestations astounded him. "No one," he wrote, "who has not lived and moved about amongst the bush of the tropics can appreciate what a torment the various . . . parasitical species are." But, imagine, he said, how bad it would be were his whole body covered with hair, and consequently, with ticks, mites, and their kin. He reasoned what a century of subsequent biology has confirmed as a rule, that the more habitat was available, the more individuals would be found. The habitat in question was the hair on his body, and at that moment, he wished there were less of it. Discomfort, if not the mother of theory, is a reasonable stand-in. This parasite theory was suggested again in 1999 by Markus Rantala, a biologist at the University of Turku in Finland, who (like me) spends much of his time studying ants. Rantala suggested an idea nearly identical, albeit more formally developed, to that of Belt. The idea was then suggested yet again by Mark Pagel and a colleague in 2004, who had discovered Belt's work but were unaware of Rantala's paper.

I have never had fleas or lice myself, but I do have a story about genital crabs, a lousy story. Genital crabs (*Pthirus pubis*) are a kind of louse. Like other lice, they spend their lives on bodies. They do not do well anywhere else, and in fact, their survival depends upon staying on their host. They are larger than body or head lice and look, give or take, like miniature versions of Ganesh, the many-armed Hindu elephant god. Unlike Ganesh, they are thin and delicate, so much so that they dry out and die when away from a host for even a few minutes. Genital crabs cement their eggs onto their host's hair, feed on their host, and otherwise go nowhere else, except in those moments of intimacy when two hosts are close enough for the lice to jump from one host to another. Our genital lice are most closely related to those of a gorilla's, suggesting that our ancestors and the ancestors of a gorilla once "interacted." It may seem that depending on a host to touch another host is an unlikely way to survive, but it turns out that touching each other is one of the things we most predictably do. Genital lice have traveled the world at our expense. Like head lice, they rode on the first peoples to migrate into the

New World. Head lice and genital lice have both been recovered from Peruvian mummies. Such is the indelicateness of death.

The first genital louse I saw was among a series in the insect collection at the University of Connecticut. It was alongside another specimen reported to have been collected from a toilet seat recently used by Dennis Leston, a famous, or perhaps infamous, ant biologist who died only recently. Leston was well-known for studying the ways in which ants in the canopies of trees can help (or hinder) pest control in orchards as a function of their identity. Some ant species eat thousands of pests; others farm those same pests for their honeydew and in the process actually increase their numbers. Leston, though, was also known for other things, so much so that in Ghana, where he long studied the role of ants in coffee plantations, locals wrote a popular song about him, something along the lines of "This white dude is not so cool." It was this lack of coolness that resulted in Leston's eventual expulsion from the University of Connecticut and also, one suspects, the louse in question. The point of the story, though, is that in order for Leston to rid himself of the genital lice he had "collected," he would have needed to shave—and let me use a euphemism here—his "louse hotel." In fact, shaving oneself is the most effective treatment for nearly every external parasite that afflicts humans, be it flea, louse, or something more exotic. In one study, doctors found that the rising commonness of "bikini waxes" was associated with a decline in genital lice, even as cases of gonorrhea and chlamydia are on the rise.[1]

Ectoparasites ("ecto" means on our bodies, as opposed to endoparasites such as intestinal worms) have an innate fondness for living on, in, and among hair. It is what makes outbreaks of head lice so common among children, and so difficult to stop. The nits cling to hair for life, as do their parents, the actual lice. Lice have special hands that are curled in the shape of the hair around which they grip. The size of their grippers is matched to the width of the hair among which they live. Head lice (and body lice, to which we will return) have narrow grippers, for narrow hair. Genital lice have wider grippers, one of the reasons that one of the few other places you can get genital lice is on your eyelashes, which are also a little thicker than head hair.

Given the intimate links between our parasites and our hair, an association between the amount of hair or fur that we have and the number of ectoparasites we feed seems reasonable. Yet this parasite theory for our nakedness still needs a bit of explanation. For as much as many of us show disdain for the fuzzy muzzle of our ancestors, the loss of our hair also meant the loss of its many advantages. Hairlessness makes humans more vulnerable to UV radiation. It makes it harder for us to stay warm without using clothing.[2] It also makes us look smaller, the way a naked mole rat looks small but a husky dog never does unless it happens to be shaved.

Genetically, it can require very few changes to cause an animal to have no hair, perhaps no more than a single change in a single gene. In general, losing a trait is easy, which is one of the reasons that we have now produced a variety of "naked" domestic animals, including dogs and cats, but also, even, chickens. That natural selection has produced very few bald mammals (and not a single species of featherless bird) suggests that being covered is nearly always useful. Canopy mammals have fur. Nearly all underground mammals have fur. Even most swimming mammals have fur. Fur is good and wonderful. What is required for its loss is a set of conditions under which fur becomes costly, whether because the hairless mate with greater success, or the hairy, bless their hearts, are far more likely to die.

The first question we must ask of the parasite theory is whether losing parasites was so much of a benefit that it was worth years of sunburn on the beach, shivering in the snow, and awkward moments in front of oh so many mirrors. What the parasite theory relies on is not that parasites themselves do us so wrong. The bites of fleas are itchy, but, in and of themselves, otherwise innocuous (in this they are similar to some of our gut parasites) except in those cases where they reach very great densities. They bite us, suck a little blood or eat some dead skin, and then go on their way. Occasionally chimpanzees and gorillas are infected by so many parasites that they develop sores, as presumably were our ancestors. Infections from these sores might lead to mortality, but not very often. It is the diseases that such parasites transmit that kill. Ticks transmit spotted fever, encephalitis, typhus, Kyasanur Forest disease, ehrlichiosis, Lyme disease, and Astrakhan fever, to name a few. Lice transmit relapsing fever and typhus. Fleas transmit the plague. To the extent that our parasites carry such diseases, losing our hair may have increased our chances of living longer, or at least long enough to mate. It is even possible that hair favors the spread of some diseases that do not require vectors. Bacteria can also live in hair (or feathers), which is why, when trying to produce germ-free guinea pigs, James Reyniers, whose story I told in earlier chapters, shaved the mother guinea pigs. It may also be why the birds that feed on dead animals have evolved bald heads three times independently, once in New World vultures, once in Old World vultures (which are actually descendants of storks), and a third time in the ancestors of the bald-headed and ungainly marabou stork.

The question that Charles Darwin posed in this context was why humans abandoned their fur but other mammal species did not. Surely, he thought, if hairiness predisposed us to parasites and their diseases, it would do the same for other mammals. Wouldn't a naked bear, as funny as it might look, suffer fewer fleas? Yet we see no naked bears, nor for that matter even any bears with sparse hair. The answer to this naked-bear paradox may have to do with two features that distinguished early human societies from, say, a group of bears.

First, even though early humans are typically described as "nomadic," they lived in relatively sedentary groups for large parts of the year. Within those groups, parasites could build up to great densities. And at the end of the day we had a sleeping spot, a spot to which a group of us would have returned. More often than not, that spot would have been a cave. It is known that by living in caves in those early days, we came into contact with bat bugs, insects that feed on the blood of bats while the bats sleep. One lineage of bat bugs jumped ship onto early humans and became bedbugs. For this to occur, we would have had to be rather predictable cave dwellers, predictable enough that the bedbugs could sleep during the day in our primitive bedding and find us again each night. Being sedentary meant that those parasites that do not pass from one body to the next could still find us. Bedbugs could wait for us in our sleeping spaces without evolving unique tricks for actually hanging onto our bodies. We also know today that animals that live in groups, especially groups that return to the same sleeping sites, such as seabirds in their rookeries, or bats in their caves, have many more parasites than do those animals who go it alone. It was this lifestyle that would lead us to pick up fleas (and the diseases they carry), even though no other primates host fleas. Perhaps the key was our living together and in particular the densities at which we came to do so.

The second thing that is different about humans, aside from our tendencies, at least historically, to live in ways that favored ectoparasites, is that we invented clothing. The invention of clothing and the loss of hair may have been roughly contemporaneous. Once we had the ability to change our temperature (and also our level of protection from the environment more generally), it may be that the benefits of hair disappeared, and so evolution had only the costs to deal with. If a trait has more benefits than costs, it will tend to stick around. If it has only costs, it should disappear. In other words, once we could make our own (washable!) clothing out of the fur of others, whether that was 200,000 years ago or even earlier, the flea-bitten fuzziness of our ancestors was nothing but a drag.

In this context, it would be nice to be able to compare our bodies and our parasites to those of other species of ancestors in those years before they lost their hair. Those ancestors should have had a greater incidence of diseases transmitted by fleas, lice, and other parasites. But we cannot make such a comparison, not yet, anyway. The closest relatives we have today—apes, chimps, and orangutans—are very different from our ancestors. Yet it seems telling that apes tend to have, in addition to the parasites we face, a broader suite of parasites including, for example, fur mites, that we now lack. Whether our immediate ancestors also faced these parasites and their pathogens is unknown, though it seems likely.

If we turn to more recent history, a few anecdotes seem telling. In June of 1812 Napoleon Bonaparte assembled his troops to try to occupy Russia via Poland. Napoleon's ambition could scarcely have been more grand. Yet sometimes ambition is not enough. As is often noted, more than half a million of Napoleon's soldiers (nearly five out of every six of his men) would die trying to control Russia. What is less often noted is that most of those troops died not from battle itself but instead from disease. They died of a spotted fever spread by lice or of dysentery. These deaths began long before the French had even seen any Russians. Only 40,000 of Napoleon's troops survived. He brought a city of soldiers and came home with a small town. The Russians, on the other hand, suffered no such fate. But why? One of the differences might have been hair. The French wore hairpieces and in doing so augmented the habitat available for lice and the diseases they carried. The Russians did not wear hairpieces. Relatively speaking, they were more hairless and as a consequence, saved. Nor was this the only example in which ectoparasites played a significant role. By some estimates, World War II was the first war in which more soldiers died in combat than of ectoparasite-transmitted diseases.

But history is not the only other example of the effects of the relationships between society, hair, parasites, and disease. As with other questions of our history, we might learn the most by looking to other species. This is the comparative approach to ecology and evolution. One can also examine other species that, like us, have made the transition to large, relatively sedentary, societies. We could turn again to the ants, bees, wasps, and termites, but we do not have to go so far afield on the evolutionary tree. Huddled with us on our own mammalian branch are the mole rats. Many kinds of mole rats live in Africa. In general, they are tuber eaters and live either wholly or mostly subterranean lives. Some species have queens and workers, as do ants. Several species have lost their eyesight. Only one species, however, is hairless. That species, like humans, lives in conditions that present a near constant environment. Once it did not have to worry about staying warm, the costs of living in a society with fur may have outweighed the benefits and it, like us, lost its fur. The difference between mole rats and humans is that there are other living mole rats with hair, so we can compare their parasite loads to those of the naked variety. Naked mole rats are not known to have ectoparasites. In contrast, all of the other mole rats so far sampled have tunnels filled with ectoparasites. This may have been what our ancestors looked like—hairy mole rats—each one of them itchy with bites and, from some of those bites, diseased.

Maybe we are naked because of lice, mites, and flies (fleas are unlikely to have played a role in our story, since fleas spread relatively recently from the New World, and plague is a relatively new disease). And maybe, just maybe, that is the same reason naked mole rats are naked too. Like many aspects of our bodies and their origins, no one is totally certain yet. Other explanations are possible. Whatever the answer, we seem more likely to understand it by understanding what happens when other mammals become hairless rather than by looking for more clues in the fossils of our own most recent kin. In the end, all of this discussion of hairlessness would be a bit silly if hairlessness were not one of our most defining traits. Once we were naked, many other things about our biology also had to change. We evolved special sebaceous glands to deal with the reality that we now had so much exposed skin, skin that needed to somehow be cooled in the hot sun. We began to view nudity with titillation (in certain Papua New Guinean tribes, only a gourd is worn by men, and yet going "gourdless" still causes a stir). The act of revealing our nakedness became the basis of a $100 billion global pornography industry. Our skin turned dark to protect our flesh and then, in some peoples, turned pale again. That paleness is the basis of thousands of deaths a year owing to skin cancer and, at the opposite extreme, the darkness of melanin is at the root of thousands of cases of rickets. Our nakedness frames who we are and how we act toward each other. Our nakedness is central, and to the extent that they may be implicated, then so too are lice, ticks, flies, and the rest. So too are the pathogens that ride in their guts and mouths, pathogens that, although small, may have been powerful enough in their influence to depilate us, one death at a time.

Meanwhile, we spend millions to make sure we maintain a little fur on the tops of our heads, and millions more to remove the fur on our bottoms. We are the naked, but high-maintenance, ape and it may be because of our pathogens and the diseases they cause. Were other hairier hominids still around, they might look at us the way we look at naked mole rats or vultures, as just a little bit disgusting. Our diseases have marked us. Long ago they shaped our immune systems. More recently, they may have made us hairless. And these are just the most obvious of the ways we have evolved to respond to our plagues.

Whatever the early influence of parasites and diseases on our hairiness, it was not the end of the story. With the dawn of agriculture, of cows and corn, things got worse. New diseases also arose and began to accumulate. Human malaria (*Plasmodium falciparum*) evolved at about the time the first crops were domesticated.[3] When it did, it appears to have spread rapidly. Mosquitoes bred in the damp spots in farm fields, and from those temporary refuges, ferried the malaria parasite from one farmer to the next, rain or shine. Once in a human, the malaria parasite takes up residence inside his or her red blood cells. Many of the first humans who contracted malaria, perhaps even most, died. But some survived, and those individuals tended to have genes that made them more resistant to malaria. One of those sets of genes, the one often discussed in introductory biology classes, can confer malaria resistance but also causes sickle cell anemia. When a child receives a copy of this gene from just her mother or just her father, she is immune to malaria and far more likely to live long enough to raise her own children. But when those genes are received from both parents, they produce fatal sickle cell anemia. Malaria is so prevalent in much of the world (particularly in tropical Africa and Asia) that these genes are still favored, despite their consequences. Amazingly, this gene is not the only or even the most common one to have helped to buffer us from the deadly tides of malaria.

The most common gene variant for malaria resistance is not related in any way to sickle cell anemia. Its name is G6PD, and it leads to the production of blood cells that starve the malaria parasite of oxygen. These are tough, protist-choking genes, malaria-killers, evidence of the power of evolution and the adaptability of man. Sarah Tishkoff—the geneticist at the University of Maryland who discovered the repeated origins of the genes for digesting milk as an adult—has recently studied the spread of G6PD. More than 400 million people in Africa, the Middle East, and the Mediterranean have one of several versions of this gene. It appears to have spread quickly, mother and father to child. These gene variants, though, like the ones associated with sickle cell anemia, come with a price. Individuals with the malaria-killing version of the gene develop anemia when they eat fava beans. Where malaria is present, surviving malaria but having to avoid fava beans is no great tragedy. Malaria still kills several million people a year, and it is therefore likely that these genes are still favored in many parts of the world, fava beans or not. But malaria is a tropical disease today, and so as individuals with the malaria-killing gene have spread around the world, their genes have spread beyond malaria and beyond their utility. Consequently, many millions of people remain unable to eat fava beans, even though they are unlikely to contract malaria. Those individuals (and you might be one of them) have genes that are no longer, in their modern context, useful. They also tend to live precisely in those areas where diets rich in fava beans are most common. The religious sometimes wonder if their gods have a sense of humor. Favism seems to be an indication that natural selection does, and it is dark and mean. Whether it is worms, crops, or disease, the more we look, the more who we were and how we lived bumps up against who we are.[4]

Our hairlessness, sickle cell anemia, and favism may all be the result of the presence of diseases. But what happens when we remove some of the worst infectious diseases from our lives, what effect has that had? We pulled the worms out of our guts and the predators out of our grasses, but what about when we removed our infectious diseases, whether through disease control or simply moving. What then? We would like to think that the answer was simply that we became healthier, longer lived, and otherwise unaltered. Unfortunately, nature is seldom so simple.

14

How the Pathogens That Made Us Naked Also Made Us Xenophobic, Collectivist, and Disgusted

When Randy Thornhill looks down on his hometown of Albuquerque, New Mexico, from the surrounding mountains, he feels as though he has to confront what the rest of us ignore. We tend to feel as though we are in control of each action we take. From Thornhill's perspective, it is less clear-cut. Our lives are more like boats being steered with loose rudders. They lurch this way or that, drawn into the troughs and swells of ancient urges.

As a scientist, Thornhill studies one organism and sees in its stories generalities that apply to other organisms. Thornhill focused for years on scorpionflies (so named because of the supersized genitals of the males, so large they resemble a scorpion's sting) and other insects, such as water striders. But even in the beginning stages of this work, he recognized in their primitive decisions the wants and decisions of his fellow humans. He sees our most terrible vulgarities and greatest delights as deriving from our evolution. Reason, to Thornhill, is what keeps us from drowning in our instincts, but we exercise it imperfectly, treading water in a sea of great waves and imponderable depth.

In 1983, Thornhill became well-known for what is now a classic book in entomology, *The Evolution of Insect Mating Systems*, written with John Alcock. The book was the original treatise on the sex lives of the smaller majority, informed by scientific observations of insects and their many ways of coming together, be it in the sand, midair, on logs, or even underwater.[1] It was not until 2000 that Thornhill became infamous. In that year he published a book in which he used models that he and Alcock developed to understand rape in insects (which is common and complicated in species like bedbugs that do terrible things under your sheets) in order to understand rape in humans.[2] The book was greeted with intrigue but also outrage. The outrage might have made Thornhill shy away from his work on the hearts and urges of humans. Instead, ever the academic cowboy, he assembled other biologists around him who study humans and human behavior from an evolutionary perspective that spans from insects to humans. From this intellectual posse, many wild ideas and wide-eyed scientists have emerged. Among the most recent of them is Corey Fincher. Fincher did not mean to become the most radical thinker in Thornhill's posse, but it happened all the same.

Corey Fincher began as a graduate student at the University of New Mexico in 1999. He planned to work on rattlesnake courtship. The sex lives of rattlesnakes are elaborate and intriguing. Fincher wanted to know the details. But things would not work out the way he wanted them to, perhaps because of the inherent difficulties of working with rattlesnakes, but also because Fincher's mind had begun drifting to other topics. It is easy to get distracted in science, to wander along the millions of roads not yet taken. Fincher was distracted by disease. Disease seemed to be everywhere he looked. He wondered how it was that species escaped disease. The more he read about disease the more it seemed surprising that any animals were ever healthy. But he still had to find something tractable to study for his thesis and so he focused on water striders, following up on work that Thornhill himself had done years before. He did so with enough success to produce a master's degree. Yet he kept reading about disease. As he did, he realized that most animals, be they rattlesnakes, water striders, or monkeys, have an immune system, just as we do. But they also have what would later be called a behavioral immune system, a suite of responses to the world that make getting diseases less likely in the first place.[3] Fincher wondered whether humans had behaviors for avoiding disease, behaviors that were either subconscious or buried so deep in the norms of cultures that they had not yet been recognized as doing anything useful at all. Soon Fincher found himself beginning a PhD. This time he would go boldly. Abandoning snakes and water striders, he wanted to understand the big story of disease, history, behavior, culture, and humans. Thornhill had looked at insect sex and seen human sex. Fincher looked at water striders, on the surfaces of ponds, and saw the long history of humans fleeing pathogens and evolving, as they did, just to get away.

Fincher knew that as humans settled into permanent or semipermanent villages, the pathogens that cause infectious diseases grew more diverse and more common. When we stopped moving, the diseases started to catch up with us. Every so often a new disease would arise. By 200 years ago, despite being naked and relatively flea- and louse-free, humans collectively hosted hundreds of different kinds of pathogens, more types of pathogens, for example, than can be found on all of the Carnivore species in North America. The process is ongoing. Even now, each year more pathogens jump from their main hosts onto us. Many of these pathogens are transmitted person to person, one body to the next. The denser our populations become, the easier it is for them to spread. Yet the fact that we have persisted this long suggests that we have many ways of coping, and perhaps some of them are behavioral. Any individuals or societies that had new and effective ways to deal with new and terrible pathogens would have done better.

Evasive maneuvers could be taken. Humans could simply move. If one moved fast enough to new places, sometimes not all of the diseases caught up. Native Americans walked across the Bering Straits and in doing so were able to shake many of the worst human diseases (the diseases caught up in 1492 when Columbus and his ship full of diseased mates set sail). But one did not necessarily need to move fast. My own research has shown that both the number of different diseases in a place and how common they are (the number of cases of a given disease) are influenced strongly by climate. Cold places and dry places have fewer diseases. Malaria, as one example among hundreds, requires specific mosquitoes to move it from one body to another. All one has to do is move away from these *Anopheles* mosquitoes to escape malaria, whether it be toward the poles or to higher elevations.

There was also another option though, changing how we behave toward each other. If being social and sedentary makes us more prone to disease, changes in the ways in which we are social might have the opposite effect. We could groom disease-causing parasites off each other's bodies. It is not romantic, but neither are lice or fleas. Grooming is an old and highly functional approach to disease control. Rats, pigeons, cows, antelope, and monkeys groom. When pigeons are prevented from grooming, they grow speckled with lice. Cows prevented from cleaning themselves have four times as many ticks and six times as many lice as those left unhindered. Antelope have a specialized tooth called a "dental comb" that seems to serve no purpose other than to aid in grooming away ectoparasites (evidence of yet another case in which ectoparasites seem to have posed a cost that was significant enough to cause animal bodies to evolve).[4] Many animals groom themselves and each other even though the lost time such efforts entail is costly. Rats spend up to 30 percent of their time grooming, time that might be spent foraging or searching for mates. Howler monkeys spend about a quarter of the calories they consume swatting at flies. Clearly, grooming is a behavior that both helps to reduce parasite (and likely pathogen) loads and varies from species to species and probably place to place.

Fincher thought about other behaviors that might affect our chances of getting diseases, other behaviors wired into our brains or embedded in our cultures. Sure, we swat flies, pick lice, and move, but moving is hard and for many pathogens, once they had already accomplished their most difficult feat of arriving on our bodies, grooming came too late. One cannot groom away malaria, or any of the many diseases transmitted body to body rather than via a vector. What Fincher wondered was whether some human behaviors and cultural practices influence the probability of coming in contact with a disease in the first place, a kind of behavioral immunity. Insect societies sometimes organize into smaller groups within colonies to reduce disease transmission. Some ants assign only a few individuals to the role of undertakers, to reduce contact with the dead. In at least two species of ants, sick workers leave their nests to die alone, where they pose no risk to their sisters of passing on their disease. Fincher wondered whether humans showed some of these sorts of behaviors, even if subconsciously—whether, in other words, we were as smart as the ants. He wondered too whether the kind of behavior humans showed depended on the frequency of their exposure to disease.

A big hint came in 2004. Jason Faulkner, a graduate student in the psychology lab of Mark Schaller at the University of British Columbia, suggested that xenophobia, the fear of others, evolved to control the spread of disease. Faulkner imagined that when diseases are common, xenophobia might guard us against diseases that travel from one tribe to the next. Perhaps it is for this reason that "others" have often, across history and cultures, been described not only as scary but more specifically as dirty and disease-ridden. It was "the others" that almost always had fleas, lice, and rats. Our dislike for the others seemed like an evolutionary universal to Faulkner, one that might be stronger where disease is more prevalent in a way that, while it causes social problems, may once have saved lives. What if, he wondered, xenophobia arose as a specific and useful form of disgust, itself an emotion with no known value other than to keep us away from disease?

Fincher saw Faulkner's work and began to pull together his even bigger, more speculative, idea. Forget the water bugs, he would figure out the story of humans. He may have been too ambitious, but far be it for Thornhill to discourage him. Fincher read papers on anthropology, sociology, and, of course, insects. He latched on to the basic attributes of human culture and behavior that seemed to differ from place to place, in particular our sense of individualism. Anthropologists have long commented on the differences among cultures in the extent to which individuals act with their own interests in mind—cowboy-style—versus those of their whole clan. This difference between individualist cultures on the one hand and collectivist cultures on the other is one of the biggest differences among peoples globally, bigger even than differences of livelihood, marriage practices, or taboos. In many Amazonian groups, one's family or clan is nearly as important as one's self. In such cultures, often referred to as collectivist, the big distinction is not between one individual and another, but between one group and another. Deviations from the group's norms are frowned upon. Individual creativity and personality are regarded as unimportant or even bad. Fincher, along with his growing list of individualist Western collaborators, imagined that collectivism might emerge in response to disease prevalence, where behaving in the "traditional" ways of the group might help to reduce disease, whereas behaving individually, in ways untested by time, might have the opposite effect. Maybe individualism and all that it leads to, from Western heroes to rogue biologists or even democracy, is only possible when societies are removed from the pressures of disease.

Meanwhile, in British Columbia, Jason Faulkner's adviser, Mark Schaller, and another student Damian Murray were studying whether xenophobia, but also other behaviors such as extroversion and sexual openness, were influenced by disease. Just as for xenophobia, being introverted and sexually conservative both seemed like good ideas when socially transmitted diseases were common. Together then, Fincher, Thornhill, Schaller, and Murray imagined that the key elements of differences among cultures and individuals were nearly all related to disease. We are who we are because of disease. Or so these men, all of them individualists, born in environments with a low prevalence of most diseases, had begun to believe.

Some of the links between disease and behavior are beyond reproach. People who live in rural parts of the tropics, where lice are still a part of life, groom each other, whereas families in New Jersey or Cleveland rarely do. This difference is a result of the differences in the abundance of external parasites from place to place. No one picks lice from hair in which none exist, but what about our other behaviors, behaviors more central to our identities? Could they really differ depending on the levels of disease into which we are born? Swine flu offered a kind of lesson in what is possible. In 2009, swine flu, H1N1, emerged as a potential threat. Anyone who turned on their TV even occasionally knew to be "on alert." And so what did people do? In Mexico, people began to stop kissing as a greeting. They refused to shake hands. Elsewhere, flights were canceled, particularly flights out of infected regions. In other words, people cut off physical contact with strangers. They became, one might even say, xenophobic, clustering, like ants, into groups. They began to call for an end to flights from other countries. They did not, of course, stop hugging their children or kissing their husbands or wives. They simply avoided other people. They thought first of their collective, their most intimate, tribe.

The ways in which we responded to H1N1 were what Fincher and his colleagues proposed happened all around the world, again and again, when diseases became prevalent. Of course, biologists have many theories, and not all of them are right or testable. But what was interesting about Fincher's theory was that it *was* testable. If he was right, people from regions in which pathogens were more prevalent ought to be more xenophobic. They ought to be more protective of their own people and less inviting to their neighbors. You can imagine, though, that many other things influence the personalities of individuals within cultures.

Anthropologists can give you long lists of the idiosyncratic bits and pieces of history that might be significant. Isolation, for example, might favor xenophobia (if the arrivals from afar are less predictable and hence pose a greater risk). Scarcity of resources might make one a little more hostile to one's neighbors. In the context of such scarcity, if there were any relationship between pathogen prevalence and the behaviors of modern humans, it would be surprising.

Fincher and his colleagues wanted to test their theories by seeing whether the regions with the greatest historical prevalence of diseases were also the same ones that were the most collectivist, xenophobic, and introverted. Many surveys have been conducted across cultures that aim to understand core attributes of behavior and personality. In one of the largest, 100,000 IBM employees in countries all over the world were interviewed. The interviews included questions aimed at distinguishing between cowboys (aka individualists) and collectivists. Using the database that resulted from these interviews and others, Fincher compared the individuality scores of people around the world.[5] What they found was that in regions where deadly diseases are more common, people consistently think more about the tribe and less about their own individual fate and decisions. They are also more xenophobic. Separately, Mark Schaller also found that where diseases are more prevalent, individuals are less culturally and sexually open and less extroverted.[6] What Fincher, Schaller, and others observed were correlations. Just because two things, such as disease prevalence and personalities, show the same patterns of variation from one place to another does not mean that one causes the other. But at the very least, the patterns these scientists observed do not rule out their ideas.

On the basis of their results, Fincher, Schaller, Thornhill, Murray, Faulkner, and the scientists who work with them have begun to think they have discovered general rules of human behavior and culture. They have looked at us from a distance and claim to understand—to see us for what we are. They may be right. But one hesitates to jump to a firm conclusion too quickly. What they have discovered is an interesting pattern, a statistical relationship between pathogens and human behaviors and cultures. Just how disease affects behavior is a more difficult question, or at least it seemed to be until recently.

As he sat back in his chair in his office, thinking about disease, Mark Schaller wondered how disease influenced behavior and culture. Schaller is the son of the great mammal biologist George Schaller. And like his father, a man who has spent years chasing rare beasts, Mark Schaller likes the pursuit, albeit of ideas rather than snow leopards. He wondered, could our subconscious really measure, in some way, the level of disease to which we might be exposed? Schaller wondered if we might all have an innate ability to recognize diseased individuals and to respond to them differently. It might be an ability more finely tuned in some places than in others, or perhaps only activated when necessary. What if our brains recognize and categorize the level of disease present in our surroundings and then without ever bothering to alert our consciousness, respond to this perceived risk? At face value, the idea seemed ridiculous. But Schaller and his collaborators decided to do an experiment to test the idea anyway. The results of this experiment seem likely to change our understanding of our bodies, our selves, and our relationship to the world.

Schaller set up a computer screen in his lab on which he played images of nonstressful things such as furniture and then either a series of images related to guns and violence or a series related to disease, for example a woman coughing or the face of a smallpox victim. Would the individuals who saw scenes of diseased individuals actually respond in some subconscious bodily way to seeing the disease? Direct links exist between the response of people to stressful situations and the production of hormones such as cortisol and norepinephrine, which can, in turn, affect immune function. But could a picture of someone looking sick really affect our immune system? It was hard to imagine that our subconscious responses to disease might be as sophisticated as Schaller's idea would require them to be.

The experimental subjects were brought into the lab. Their blood was taken and then they were shown the neutral slide show and one of the two sets of stressful slide shows. After the slide shows, the participants' blood was taken again. Each blood sample was then exposed, in a test tube, to a compound found in many pathogenic bacteria, lipopolysaccharide. Schaller and his colleagues thought that the blood cells of the participants who had seen the images of disease might more aggressively attack the bacterial compound by producing more cytokines. But, truth be told, they had no idea what they would see. Then the results came in. The blood taken from the individuals who had seen the disease slide shows produced 23.6 percent more bacteria-attacking cytokines (IL-6) than did the blood taken from the same individuals before the slide shows. But what about the individuals who saw the violent slides without images of disease, perhaps the response was just due to stress? It was not. The blood of the individuals who had seen the violent slides did not change at all. Seeing signs of disease primed the participants' immune systems to respond to a pathogen like *E. coli*. This happened simply because they saw the images. It happened subconsciously. It happened incredibly quickly and easily. If you walk outside of your room and see someone coughing, it is likely it will also happen to you.[7]

What Schaller and Fincher have gone on to argue is that in addition to our immune system (and perhaps our hairlessness), we also fight disease through a behavioral immune system. That system is born in part of an emotion, disgust, which rises to our consciousness but also seems to directly affect our bodies, behaviors, and cultures. It seems possible that because of this system, in places where diseases are common, we more innately express behaviors that reduce disease risk. This may include xenophobia and other attributes of who we are. In addition though, our behaviors are modulated by culture. Collectivism, for example, and other features of societies come to be encoded in taboos and norms. Norms may be shaped by the innate biology of individuals, but they also have a life all their own. Even if disease prevalence is reduced in a region, culture is likely to be slow to change. A case in point is the correlation that Fincher, Thornhill, Schaller, and others are able to show between disease prevalence and individualism. Our behaviors and cultures seem associated not with current disease prevalence, but instead with historical prevalence, the diseases we used to face a few hundred years ago. Old habits die hard, leaving us, once again, with ghosts of our past.

How does all of this relate back to who you are today, wherever you are? It suggests that how you behave toward your friends and strangers is shaped not just, as you might hope, by your consciousness but by something deeper. The manifestations of this effect may include many aspects of our personalities and social behavior, but even if it includes nothing more than disgust, it is of consequence. Disgust evolved to trigger us to distance ourselves from disease-related stimuli and to trigger our immune systems to "get ready." But the stimuli that trigger our disgust are imperfectly honed. Our minds seem to have evolved in such a way that they make mistakes in what they judge as signs of disease, perhaps because for a long time it was better to make a mistake and avoid someone who was not diseased than to make a mistake and not avoid someone who was.

For those of us with the good fortune to live somewhere where infectious diseases are rare or, because of changes to our lives and public health, have become rare, there are many potential costs to responding to the wrong cues associated with disease (and, in the absence of many diseases, fewer costs to letting an occasional disease slip by). The most obvious cost is that our immune systems and behaviors related to defense against disease may become hyperactive. Notably, Schaller's study of the response of individuals to disease stimuli showed them pictures of sick people, pictures not unlike those we see on TV every day. Could our bodies be reacting not just to actual sick people but also to television sick people? No one knows.

A more insidious and perhaps significant cost to our bodies' misreading the signs of disease has to do with social groups that our bodies lead us to subconsciously avoid. Schaller has begun to argue (and in bits and pieces to demonstrate) that many of the attributes of old age, diseases that are not infectious (such as morbid obesity), and disabilities trigger our disgust reaction. If so, they do so by accident, our subconscious minds having mistaken the signs of age, obesity, or disability for indications of infectious disease. Schaller has shown that when individuals perceive disease to be a threat, they are more likely to act in ways that can be construed as ageist.[8] Similar results hold for our perceptions of the obese, which are worse when we are worried about getting sick. These responses, if real and general, have broad implications for how we deal with the aging, disabled, and chronically ill in our societies. That the elderly, the chronically sick, and the disabled can become marginalized in many modern societies is beyond doubt. That such marginalization is a result of our misplaced evolutionary disgust, disgust that evolved to save us from disease, is more speculative, but plausible. Regardless, it seems that our behavioral immune system, whose intricacies we have yet to fully understand, is only partially functional in the world we have made for ourselves. It tugs subconsciously at our actions and our immune systems even before we are able to consider what is right or wrong.

In the meantime, Fincher and Thornhill continue to think even more radically. They cannot help themselves. They wonder if the greater xenophobia and collectivism of cultures in high disease areas leads to boundaries between cultures and the accumulation of differences between peoples. What if, they have suggested, diseases, xenophobia, and collectivism also make democracy more difficult to attain or maintain? What if they make war more likely too? So far all of these theories find some support, but they are nascent and wild and will take more time to better understand. In every case there seem to be alternative explanations, yet these theories invite pondering, if only because, if correct, they affect nearly the entirety of human history. In the meantime, if you are intrigued, Thornhill is still taking students, but you had better be an individualist if you intend to apply.

PART VII

The Future of Human Nature

15

The Reluctant Revolutionary of Hope

If our hairy ancestors were to visit our urban and suburban lives, they would wonder how the escalators work, but also where the plants and animals have gone. What have we done with all of the birds? Of course, the answer is that many of them are geographically removed from our daily lives, far away from the majority of people. Our food is shipped to us. It comes wrapped in paper, decorated with designs and shaped by a machine. It is stamped with its contents, rather than its origin or history. Cows are milked not by hands but instead by mechanized devices. The chickens whose meat you bite into were grown indoors. Our mutualist species, living plants and animals on which we depend, have been turned into material, matter that we consume. In this regard, our cities are different from any cities that have every existed, and so too our lives are uniquely disconnected from the daily lives of the species we depend on.

What can we realistically do to restore the good elements of nature to our lives in London, Manhattan, Tokyo, or Hong Kong or, for that matter, Raleigh, Syracuse, and Albuquerque? A first answer is to recognize the importance of the structure of the city, its complex framework of buildings, dirt, roads, and pipes. The environment we create influences our interactions as much as does any individual decision we make within that environment. I began to think about the infrastructure of our lives while living in Bolivia, where many versions of the future remain possible, and where one might step back and, with sufficient appeal to reason, vision, and power, make great change. It happened at Tom's, a small restaurant in the middle of the plaza in Riberalta in the northern Amazon. It sits at a prominent corner of the middle of nowhere. In the relative world that exists in any place, it is a den of the wealthy. A series of tables are set outside, without umbrellas. At those tables, the most affluent people in town eat meat and soup. Nothing that can be ordered costs more than a few dollars, but even this modest tithe divides the world. Perhaps one in a thousand people in town can afford a lunch at Tom's. Even fewer can afford to lunch there often, and so to sit at the tables near the plaza where motorbikes circle en route to narrower paths is to announce oneself as fortunate.

I sat at one of Tom's tables on the morning I met the woman who would tell me the story of the future of cities. In her telling, her father was a city planner who had lived his life in the highlands. Bolivia is divided into highlands and lowlands, and the two have been culturally divided, to different extents, for millennia. Her father was a highland man with the kind of education that earned him a seat at most tables. He was one of the most important city planners in the country. He had been tasked with planning a new city, a visionary city. In Bolivia, vision has a history. It was in Bolivia and neighboring Peru that the Incan empire rose out of mud to grandeur. The Spanish never found the golden city, but had they broadened their view of gilded to include masterpieces of architecture and urban planning, the Incan empire would have fit the bill. Given vision, grandeur could rise again.

This man's primary task for his entire working life would be to design a city to befit the descendants of the Incas. Into this city, he drew the streets and buildings. He put in parks, pools, and apartments. He added houses and then imagined and drew the flowers to grace each garden. Hydrangeas by the wall. Roses on the hill. He added administrative buildings and plazas. He drew and redrew. Great piles of drawings were carried away each week by the garbagemen. The city was, in its stubborn way, reborn each month, year, and decade until, one day, a final draft sat on his desk, complete and wondrous.

In the process of making his final draft, the father and city planner imagined himself moving into a house with his wife. He made a restaurant on one corner, where he thought his daughter might meet a man and fall in love. It was not just that he could imagine things, but that he could, as his decisions progressed, come to control them. Making the streets narrower, he could send a bicyclist into a ditch. Widen them again and he could rescue the bicyclist and allow him to ride all the way to work. He could move benches to change where old men would sit to talk. He could point statues so as to orient their stone gestures, but also the pigeons that would come to rest on their shoulders. He could see parrots flushed from fruit trees, and fruit falling to schoolchildren's outstretched hands. He came to know this city, his city, as a kind of orchestra. He wanted to play all the instruments just right so that the whirring music of lives being lived sounded not just good but perfect.

Because he had time to choose and rechoose, the features of the city that he planned were those that appealed to universal human wants and preferences. Habits and cultures change, morals too. Passions, society, and the need to exist with other life remain. His city would be built to please people, to make them happier and healthier for centuries. The need for a dog, the flowers, and thousands of trees, like the need for places to meet and talk, would last long beyond his life. Under the surface of these imagined lives, he penciled the other necessary layers—pipes for water and waste, and the rest of the infrastructure of persistence. Some men and women have the patience to plant oak trees and wait for their shade. This man would seed the next century and then watch it begin to grow.

As I sat at Tom's, I was intrigued by this story and what it meant about the potential scale of one's dreams. But it presented me with a problem. I knew of no city like the one being described, not in Bolivia, not anywhere. Yet the daughter waved her hands and spoke of it as if it were finished. She spoke as though she had fallen in love in the restaurant on the corner, as if her father had in fact pulled the bicyclist out of the gutter by the force of his design, sent the pigeons flying, brought the singing birds to roost in the president's trees, and ripened the fruits on every orchard tree. So I asked, *"Donde esta la ciudad?"* but even as the words stumbled around in my mouth, I knew the answer. The city was never built. It would never be built. Her father had come to know that too, yet he spent nearly every day of his life in his office designing a city that he knew was only a place of imagination. In its purity and grandeur, this woman's story is a thing of beauty, a kind of architectural novel. In it, the plot is about the extent to which we have the ability to change the fates of lives, both those of other humans and of other species, but the subplot is, of course, that even when we plan, we do not always succeed. Yet we go on planning, putting pencil to paper, and dreaming.

This woman's story about her father is what separates us from the other societies. The ants may make networks of roads that are more optimal than our own. They may work with far more efficiency. They may use their resources in ways that are more sustainable, and may even live in societies in which a greater diversity of life thrives, and yet they lack what this man had, the ability to sit back and plan a vision on the basis of our collective reason. The ants lack the ability to take hold of their fate and, in doing so, to make a change.

We do not always dream or decide consciously. Many days we are rather like the ants, pushed this way and that by our urges and conditions. Collectively though, we have the ability to learn and extend beyond our individual limits. We have the ability to develop a plan and on the basis of that plan to enact change that affects not just our own lives, or even those of our own species, but instead all of our lives and all species. We have the ability to pick up the drawing pad on which the future will be laid out and sketch the streets, the houses, and the people, our descendants, moving back and forth, and to decide whether they walk or drive, but also how they interact with each other and the rest of life. What I will leave you with is not a set of answers, a how-to book for the biological future, but instead the story of a few other visionaries with pencils, individuals different from the Bolivian urban planner only because what they have sketched out may yet come to be.

Dickson Despommier did not mean to become a revolutionary, or to plan the future of a city. He just wanted to be a scientist. He grew up like many of us. He caught dragonflies off his mother's clothesline and put them in big mason jars, where he would watch them trying to will themselves out of the jar. He gathered snakes. He poked at nature and explored in the way that any child might do, trying to find, if not truth, amusement. Those days of fumbling with life led to graduate school and a postdoctoral fellowship, and then on to a career studying parasites and in particular a worm, *Trichinella spiralis*, that Despommier still finds, for lack of a better word, beautiful. It is also terrible, of course, because it causes trichinosis, but in the elaborateness of its sinister ways, it is also elegant. Fumbling with life led Despommier to this worm and so he spent twenty-seven years with it. In doing so, he learned much about its life and the life of parasites generally. He became an elder statesman of parasitology even before he was elder. He invented a rapid blood test to detect whether humans are infected with *Trichinella spiralis*.[1] All of this proceeded as he might have hoped, probably better than he might have hoped. He saved lives and had big insights. Then in 1999, at the age of fifty-nine, he found himself in a new situation. He could not get funding, not from the National Institutes of Health, the National Science Foundation, or anyone else.

Times can change and leave scientists behind. New fields gain traction, whether or not they represent progress or truth, and then the old fields disappear, either permanently or just for a while. These fields' diligent geniuses go unfunded and ignored in the defunct arenas that face the same fate. Despommier was watching the birth of the field of genomics, a particular kind of industrialized genetics, and with it the quieting of the study of how species actually live and work in the world. He applied for one more round of funding and then another and another, before finally deciding to focus on teaching. Up until that point in his career, he had not really thrown himself into teaching, but now, with time and energy (and no money) on his hands, he would. He would focus the intensity he had once reserved for discovery on the minds of the graduate students of Columbia University. They were students, on average, with a sense of entitlement, but also a high probability of future lives of consequence, and so he would turn to them.

Despommier began to teach two classes for graduate students. One was an environmental health course called Medical Ecology. The other was Ecology 101. It was while teaching Medical Ecology that his life began to change. The course proceeded reasonably, some students excited, others disinterested, some students friendly, others passive, some students sleeping, most awake. It was, in other words, a typical class, at least until Despommier made a decision that would cause him to lose control of the scope of his life.

Dickson Despommier was walking his class through the ways in which the world is collapsing. By 2050, the world is expected to be occupied by 9.2 billion people. It will be hotter and harder to farm. Diseases caused by pathogens will once again be a key problem, not just for developed countries but for the whole world, and all of these issues will coexist with our modern problems that seem to be getting worse rather than better: obesity, immune diseases, social discontent, and the extinction of thousands, maybe millions, of species. "Feeding this future world and keeping it healthy is beyond current abilities," he would tell his students. By 2050, with current farming practices, "we will need an area of additional agricultural land the size of South America. It just does not exist! Not on Earth!" Despommier said this, all of which is true to the extent that it is knowable, and the students reacted. They started complaining.*

Students do that. It is their nature, which is to say human nature, a nature of mild discontent. These students were insistent. They were sick of hearing about doom and gloom, sick of hearing about how the world into which they were maturing was falling apart. They were filled with youthful expectation and wanted, simply, but perhaps too simply, to talk about something more hopeful. They were the ones paying (or at least their parents were). It was "their class."

The natural thing would have been for Despommier to remind the students that he was the one teaching, and that much of what he had seen of the world was not particularly hopeful. He might have offered some examples of positive change, and then moved on. Or he might have retaliated and made even clearer to the students the magnitude of the doom. "You have not seen the half of it," he might have said, like the archetypical elder on the porch. But a mood struck him. He decided to look for causes for hope, or at least make the students do so. "Hope away," he thought to himself as he asked the students if they could think of a way to solve some of the problems he had, by that time, already laid out, problems that affect billions of humans and more and more each year. So it was that at an age when his colleagues had begun to talk about retirement, the seed of Dickson Despommier's new life as a hopeful revolutionary was planted. Out of it, a new future would grow.

The students were faced with the problem that faces all of us: what to do about the ecological situation in which we find ourselves, a situation in which we are removed from the kind of nature that once

---

*A hint of the course's flavor may be discerned by a quick look at newspaper articles in which Despommier was mentioned or quoted during those years. "Diseases carried by mosquitoes are a leading cause of death." "Rise in international travel helps viruses cross borders." "Contaminated food makes millions ill despite advances." "West Nile virus found in organ recipient." You get the picture.

plagued us, but also the kinds that once benefited us. They looked around for hopeful solutions and decided to study green rooftops, the patches of flowers, trees, and even crops seeded on the flat tops of city buildings. Green rooftops exist already in many urban centers, patches here and there of vegetables, grass, or other photosynthetic life seeded in the dirt on roofs or balconies. Some green rooftops develop accidentally. In tropical countries, any roof left untended for a few days will tend toward rebirth, but in other cases they are sown. Dirt is carried up steps or elevators, positioned, and then tended to as though it were any other ground. Green rooftops were, to Despommier, a boutique answer to a big, grisly problem. But he humored the students.

The idea of green rooftops and rooftop gardens is old. The hanging gardens of Babylon, if they existed at all, were a kind of rooftop garden. Nebuchadnezzar II is said to have built the gardens in 600 BC to satisfy the cravings of his wife, Amytis of Media, for the trees and plants of Persia. Clearly, rooftop gardens, whether they are in Babylon or elsewhere, have costs. Roofs must be strong and waterproof, and often water must be carried or pumped up. The gardens though also accrue benefits beyond those of simply producing food and pleasing spouses. They sequester toxins from the air. They filter and collect storm water and reduce sewer overflow. They reduce building heating and cooling costs. At the scale of cities, they have been argued to reduce temperatures on hot days. Then, of course, they make us happy. For Amytis of Media that happiness was being reminded of her home in what is now Iran. For us, today, it is being reminded of those billions of years that we lived in the wild.

Recently, green rooftops have increased in popularity. In 2008, the number of green rooftops in North America increased 35 percent relative to the year before, and this growth appears to be continuing. Even New Yorkers, who are particular about the aesthetic of their city, an aesthetic that tends toward shades of black and gray over avocado, do not seem to mind them. A rough-and-tumble tenement building on the Lower East Side is now green on top.[2] Pace

University is seeding grass over shingles. In Chicago, City Hall is now green, as are hundreds of other roofs, millions of square feet. As you fly over cities, you might see them, the patches of leaves, pockets of life where little might be expected, rising out of the gray patchwork of the modern world.

The benefits of rooftops extend not just to humans, but also to other species. Green rooftops offer testimony to the extent to which nature, or at least some of it, abhors a vacuum and overcomes a distance. Despite being suspended in midair, the brown patches on rooftops quickly fill with life, whether or not anyone tends to them. Seeds arrive along with hundreds of bee and wasp species. Spiders extend loops of silk with which they ride. Nor is it just that species colonize these spaces. They move among them, the way they might move among fields of tall grass. Winged animals fly from building to building unaware that anything they are doing is unusual. In Japan and New York, beekeepers depend on such rooftop life for the production of honey. The bees fly among buildings gathering nectar that they bring back to provision both their brood and (unintentionally) their keepers. That green rooftops become ecosystems is beyond question, but that gardens on roofs could actually prove useful at a scale that mattered to humans was a separate matter, one about which Despommier would continue to hold his tongue.

When the students began, they knew they faced a challenge even if Despommier did not explicitly discourage them. It is one thing for rooftop gardens to be a beautiful and interesting element of a city, living gems scattered here and there. It is quite another to feed thousands or even millions of people. For Christians, Jesus fed the masses out of a single loaf of bread. The students wanted to do it out of asphalt and sky, if only to keep Despommier from telling them how bad things were. No one had really "gone big" when it came to green rooftops. No one was seriously advocating for whole cities of green roofs. Perhaps, the students began to think, it was just that no one knew how valuable they might be in remedying pollution and producing crops. So they decided they would figure out how great a role such rooftops might play if they really caught on. What scale of problems could they really solve? The students thought that the answer must be, whatever it was, a large number.

The students worked hard. In a time before Google, their first step was to travel to a basement—the map room of the New York Public Library—to figure out the surface area of the roofs of Manhattan. They needed to know how much roof could be farmed. As they gathered and measured roof after roof, the answer seemed as though it was a great deal. They were not just finding roofs—there were also balconies, abandoned lots, and old railway lines. The city, for all its modernity, was filled with layers and levels of dirt, and so too the possibility of layers and levels of life. The students came back to class, week after week, ever more excited, until they found the total amount of farm that was possible in Manhattan.

The students summed and resummed. They calculated pounds of crops. What they were not able to do was turn those pounds of crops and other benefits of rooftop gardens into dollar values. A recent three-year million-dollar Toronto study, however, provided help. The study by the Department of Architectural Science at Ryerson University found that if all of the rooftops in Toronto were greened, the economic value would be substantial. Storm-water capture would lead to a net benefit initially of $118 million. Reduction in sewer overflow would save an additional $46 million. In cold Toronto and probably also in cold New York, tens of millions more would be saved in energy costs. All told, Toronto's savings was roughly $300 million initially, and then tens of millions of dollars each year.[3] Although no one had calculated it, the much greater surface area of New York's buildings would produce, undoubtedly, a much bigger savings. This was all on top of the benefits of actually producing food.

But then Despommier, still with a bit of realism's doom in his heart, asked, "How many of the people in New York City would it feed? How many of these eight million individual primates could eat from the city's vines and trees?" The answer they were looking for was a big number, 3 million or even more, maybe all of the city. The reality was humbling. It was just 2 percent, 2 percent of the food of the city. Two percent was meager—an organic mango where what was needed was miles of grains.

Despommier could have left the students to wallow in the realities of the magnitudes, the big numbers of demand and the small numbers of supply. Something compelled him, though, to do otherwise. He asked a new question. "What if we turned whole buildings into farms? What if we used hydroponics and made abandoned buildings into biomes and grew vertical farms, up walls or even inside walls, in the way that forests grow vertically?" The students were still high on hope, and so this sort of nudge was all they would need to take a wild leap.

Until this point, Despommier had played a passive role in the students' endeavors. He helped and guided, but the students' obsession was not his, not at all. He had another class to think about and he was still, privately, imagining all of the projects he still wanted to do on worms. If the National Science Foundation had funded him at that point, he might even have abandoned his class midstream. They did not, and so he kept at the class, more involved at each stage, until he found himself sucked into the students' questions. He found himself actually wondering, as he talked to his wife across the dinner table, what one could do with a little hope and some seeds.

Despommier had begun to think about the project all the time. He looked up at the buildings around him in Manhattan. The buildings were filled with human bodies and the species that lived off of them—worms, mites, bacteria, and flies—but they did not give back, not life anyway. They just took. Each day, thousands of pounds of food and millions of gallons of water were shipped and carried up elevators, staircases, and pipes, and near-equal amounts of waste were shipped down toilets. Each building sucked the juices out of the land outside the city, sucked at the land around the world. This was what he knew and had known—the gloom of the world, the dark and foreboding realities of our condition, the beaten down farm fields, the plowed-under forests, and the poor farmers from India to Brazil. He knew this and yet ideas turned through his head like windmills. Here he was, caught up in the dreams of students, and all he could think to do, rather than slowing down, was to raise his wooden sword and charge.

What would one really need to do to grow food in cities, lots of food, and in doing so make them like any other ecosystem that produces instead of simply taking? He began to make sketches on napkins. He was done submitting grants for his science on worms. His life was changing under him—the landscape he had known for all his professional life was substituted for something else entirely, a terrain as new and different as the canals once thought to exist on Mars. Hopefully, it was not also as illusory.

There were models for what Despommier and the students were working on (other than Don Quixote), but not many. There was the landscape architect Fredrick Olmsted, who came back from the Civil War and built the greatest parks of the United States in Chicago, New York, New Jersey, and elsewhere. Olmsted, though, was different from Despommier. He built the parks as public goods, paid for with public money. He built them in such a way as to appeal purely and directly to our ancestral preferences for grass, a smattering of grovelike trees and water, but his parks did not make money. Nor did they feed people. Like museums, they relied on public funds forever. Someone will always have to pay to trim the trees and keep the paths open for walking. The grass too needs to be mowed, a task that was once accomplished by one of our mutualists, sheep, but that eventually gave way to machines and the price of their gas. This model was not quite what Despommier wanted, though it was recognizable in its grandness. After all, that Olmsted was able to transform entire cities suggested that it was possible. It suggested that a single woman or man with a plan could shape the interactions of people with each other and the rest of life for tens, hundreds, or even thousands of years.

The other models came from other animals rather than architecture, agriculture, or design. Leaf-cutter ants have grown entirely reliant on their farmed food, without alternative, fallback, or recourse. The same is true for leaf-farming termites and the many fungus-farming beetles. None of these societies can return to hunting or gathering any more than we can. They are dependent on the fungus that grows on the leaves they bring back to their nest or the wood in which they tunnel. They feed it to their larvae and nymphs, scattered in colonies throughout the tropics, one small white clump at a time. What is interesting about the farming by leaf-cutter ants, termites, and all the animals that farm successfully is that they do it where they live, at the heart of their cities. They farm in the realm in which they have the most control and in which pathogens are kept most easily at bay. One can speculate as to the specific evolutionary forces that created this scenario. It is hard to imagine, in a way, an ant species ever farming a fungus outside of its nest for the very reasons that our own farms face such difficulties. Outside the nest, pathogens abound. Outside of the nest, the distance between food and child is greater. Outside of the nest, it gets too hot or too cold. Inside the nest, the fungus can be tended to, nursed the way one might a chosen child. Suffice it to say that for some or all of these reasons, every time farming has evolved, species have farmed where they live—every time except in our own societies. We are the only animals that farm far from where we live. In doing so, we have divided our world into the places that produce and, separately (in our cities), the places where we consume food and make waste. No animals have ever chosen to live beside their waste rather than beside their food. If there were ever any ants, beetles, or termites that ran their cities in the way we do, they are gone. They became extinct.

What the class and Despommier would come to propose was a thirty-story tower, or in the parlance of ant nests, a garden chamber. The tower would be the beginning of what would someday be a series of towers that would produce fruit, veggies, grains, and all the rest. It was to be a tower of food, much like the leaf-cutter ants' and leaf-farming termites' towers of fungi. Despommier and his class would design these buildings in such a way that they cleaned wastewater, generated energy, and provided other services to society. Despommier and the class estimated that 150 of these buildings could provide sustenance to all of New York City, a city in which there are many more than 150 abandoned buildings being used for nothing at all.

The math was crude, the ideas coarse, and yet the amount of food that such a building seemed capable of producing was astounding. The more Despommier thought about it, particularly the more he thought about in the context of the doom and gloom he was used to, the more it seemed to make sense, and the more it seemed as though it would make even more sense in the future. Despommier knew that by 2050, the earth will house no fewer than 2 billion more people than it houses today, 2 billion people for whom we must find plots of land to produce food. Doing more of what we are doing today will not feed those people, not without cutting down most of the remaining forest on Earth. At the same time, most of those humans will be in cities, and so we need that food in cities. We also need forests and grasslands, wherever they might be. We need them for many reasons, but especially now to remove some of the carbon dioxide from the air, carbon dioxide produced in no small part by our traffic of food from where it is grown to where we consume it. Here then, at least on paper, was a kind of panacea.

The idea was not totally without precedent. Crops have been grown in buildings for years, particularly in water (rather than soil) as hydroponics. As a *New York Magazine* article pointed out in one particularly telling example, a family in Florida farmed strawberries on a thirty-acre farm in Florida.[4] They farmed strawberries there until Hurricane Andrew wiped them out. After the hurricane, they started fresh by growing strawberries indoors in a hydroponic system, in which the strawberries were stacked one layer on top of another. Indoors, they could grow the same quantity of strawberries on one acre that they used to grow on thirty. One thirtieth of the land was occupied in their new endeavor and the rest, from all appearances, lay fallow, each acre returning, slowly to forests, birds, and bees.

Farming an entire thirty-story building hydroponically was a grand and, in some ways, quixotic vision. The popular press immediately became caught up in the mood of Despommier's ideas and what, at some point during this journey, had become his hopefulness. Each time a new article ran, more people wrote in to say how excited they were. There were always cynics. The devil, they would claim, is in the details. The critics made some reasonable points. The buildings in major cities would have to compete for land with commercial properties. The land would be expensive. Yet it seemed possible to build a green building, somewhere, on some scale and maybe many places on many scales, maybe.

Two things seemed to go missing in the discussion of the vision of Despommier and his students. The first was that Despommier did not really know what he was doing. He was a worm guy who also worked in farms, but the idea of designing an entire building, much less one in a major metropolitan area, was forty steps beyond what he really knew. The other was that what Despommier was proposing offered resolution not just to the problem of food, but, at least in part, to many of the problems humans face. They were the problems associated with a disconnect between our modern lives and the ones we once lived, but more specifically our disconnect with other species. In essence, what Despommier had proposed was to consciously build back into our lives one set of those species that benefit us, our food plants. He would do so while keeping at bay those species that would do us harm, or at least those that would do our crops harm, by creating his farms indoors. In theory, no pesticides would even be necessary. Despommier stopped at farming food. That was enough of a vision for one man, but when his vision is combined with those of others, one can look forward to an urban landscape in which key elements of our lives are resurrected around us.

Despommier's vision, though, was still that of a worm man dreaming big. He did not know architecture. He did not know plumbing. He did not know solar power. He did not know any of what he would need to know to make this new dream more tangible. Then big architectural firms started to call him and help with the logistics of designs. The designs have improved, and logistical necessities have been tended to. What was more, other people started to run with his ideas. The Internet is now populated with hundreds of images of plans for vertical gardens or farms, bits and pieces of which appear to be popping up in actual buildings, if not yet at the scale Despommier envisioned. Finally, Despommier was invited to meet the mayor of Newark, New Jersey. New Jersey, for all of its downsides, is the garden state, though it would be hard to call Newark the garden city. The mayor wanted to make changes. The worm man whose name means "of the apple tree (*Des pommier*)," had an idea that just might take seed.

Before we come back to the meeting in Newark, it seems worth revisiting not just where we are ecologically, but where we will go if we continue with business as usual. As I hope I have convinced you by now, business as usual for the past hundred thousand years has meant that we kill off what we can, farm what best suits our taste buds, and then accidentally favor the sneaky species that survive despite our best intentions. With newer tools, what we can kill grows more diverse and smaller. So too the sneaky go from being rats, to rats and antibiotic-resistant bacteria. This has created our modern urban situation, one in which the only species that are most abundant are those that persist despite us (rats, pigeons, drug-resistant bacteria, drug-resistant roaches, and bedbugs). In old cities, the biggest areas of wild nature are uncleaned alleys and the waste piles of a variety of sorts, where a kind of wildness reigns. In other words, the species present around us are there for reasons other than our active planning. The species we manage because we like them live far away. Business as usual will push the crops to more remote places and the remaining life out of cities. This has already happened in some new cities in China, Brazil, and elsewhere.

It is sometimes suggested that what we need to do in cities is to restore "nature." A body of literature and theory often referred to as "biophilia" posits that we have an innate fondness for nature, and so restoring nature to our lives makes us happier and healthier. I disagree for what might seem to be (but is not) a subtle reason. Namely, by any reasonable definition, the species that have filled in around us in cities are nature. The species that live on our bodies are also nature, as are both smallpox and toucans. What is missing from our lives is not nature, but instead a kind of nature that most benefits us. By that same token, the life we love to have in our neighborhoods and daily lives (*bio* = life, *philia* = to love) is not all life, but the life that benefits us in some way. When tigers chased us, we had no innate love for them. When diseases killed our families, we had no love for them either. And so what we need in our cities and suburbs as we move forward is not simply "more nature." More rats would be more nature, as would more roaches and mosquito-vectored diseases. No, what we need are more of some aspects of nature, its richness and variety and, more pointedly, its benefits.

When we think about benefits, we cannot think simply about what our eyes tell us. The species that benefit us in the future may well include worms, ants, and our gut microbes. They may include, in other words, a much more burgeoning ark of life than we tend to consider when we plan parks and gardens. In our guts, we may really need to give ourselves worms. That we hesitate at this point is largely a function of what our eyes tell us, not of clinical results. We do not yet understand the way worms work as a treatment, but neither do we understand the ways in which most of our modern medicines work. Ask a researcher how Ritalin or pain medication works. In most cases, no one knows. We just know that when taken, a symptom or even a disease goes away. So it is, for now, with the worms.

Inside our guts, we may be able to manage for specific bacterial species. We may be able to take probiotics that help the bacteria that benefit us, and disfavor those that would do us harm (or disfavor, in any case, the conditions under which such microbes tend to do us harm). As of now, the evidence for the benefits of one particular probiotic over another is ambiguous. Time will provide more nuance. Intellectually, it seems likely that when the composition of our gut fauna comes to deviate from that of a healthy gut, we ought to be able to manage the species of our guts in order to make ourselves healthy. Practically speaking, we are not there yet. In the meantime, the IgA antibodies and our appendix go on fighting the good fight to keep things working as they long have worked.

In our brains, we still feel as though predators lurk nearby. Fear drums our hearts. It raises our levels of stress. This fear seems to be at the root of some of our psychological ill health. We cannot reintroduce predators into our brains in the way we add worms back into our guts. Instead, what we have tended to do is medicate ourselves. We have taken antianxiety pills by the billions. Such pills may have negative effects, but in the short run they tell the brain in essence that "the cougar is not there," and so the brain rests.

In addition to giving ourselves worms and probiotics, we might manage the living world around us to be richer and more diverse with the kinds of interactions we once had. We can bring, if not wilderness, wildness back into our lives. Of course, maybe I am just caught up in Despommier's chase. Maybe he is talking about organic mangos for Whole Foods, not wheat for the masses. Maybe, one might say, at least until Despommier started talking to the mayor of Newark. Sometimes when one chases windmills, one catches them and finds that they can produce most of the energy for a city of people. So too sometimes when one chases a wild building filled with oats, it actually comes to be built.

Despommier went to Newark unsure what would result. Investors were joining him and the mayor, Corey Booker, the man who wants to put Newark on the map for something other than its smell. Despommier brought pictures of his futuristic plans for vertical gardens. He brought the statistics about the number of people on Earth, dwindling food supplies, and even more dwindling forests. He stood in the front of the room, and he pitched his idea as if our future depended on it. He was now, officially, the one who was dreaming big, who was hoping for the future. The mayor's people asked cynical questions about one part of the project and then another. The investors asked questions. Then after more discussion, the mayor and Despommier had a private meeting to talk about the future.

In the private meeting with Despommier, the mayor's people agreed to move forward. They would have a prototype built, a small vertical garden. Let us see, they said in essence, how many people we can really feed. A prototype garden is a tiny germ of what Despommier believes to be possible, but then again one might say the same of an apple relative to its tree. Meanwhile, another garden building is in the works in Italy, and elsewhere conversations are beginning, conversations where individuals give in to the power of what ants lack, the power to dream.

In the meantime, the future of urban life is not hitched exclusively to Despommier's tractor. Rooftops are being greened whether or not vertical farms are ever built. Rare species are being restored and as this happens, more and more of us can sink our hands into that life. We know some of the reasons we need other species. As time goes on, we will learn others. More remains unknown about our bodies, and so there is much to explore and understand, as there will be for generations. Given the choice of which and how many species to live around (a choice we still have, though not forever), why not consciously garden a greater diversity of life around us, be that life our crops or something even richer? Why not foster the conservation of the interplay between humans and the rest of life, to paraphrase René Dubos, and favor the species from which we benefit, rather than those that haunt our cupboards and the walls of our houses. Why not? We hang bird feeders to favor beautiful animals. Why not design whole cities to favor the species that tend not just to our pleasure, but also to our sustenance and sanity?

The question is which species we might favor, and how. Favoring our food crops, as Despommier has suggested, might be a good start, as would favoring the other species on which they depend, their pollinators, for example. Such species benefit us directly. We eat them and eat of their success. We could have honeybees, bumblebees, hummingbirds, sunbirds, and other curve-beaked nectar-lovers in our cities, but we could also have more.

To answer the question of just which species we could favor, it is worth revisiting our own story, the one that begins with the first life, passes through our potential ancestor Ardi, and leaves off in our homes today. If I retell this story in light of the material we have covered in this book, it goes something like this. Once upon a time, we lived life in nature's tangled bank. Five hundred million years ago, our hearts evolved to pump blood. Their beating was physiological, but nearly every subsequent elaboration on our bodies related to interacting with the rest of life. Four hundred and ninety million years ago, the first eyes evolved in order to detect prey. Later, the first taste buds evolved in order to help us find our food species and avoid toxic species, to urge us toward what we needed and away from what we did not. Our immune systems evolved to detect microscopic creatures and distinguish among them, favoring some and disfavoring others. All of this we share with most of the rest of animal life. In this way, our bodies unite us.

As we moved closer to becoming human, a few traits became especially accentuated. Our vision improved, whether to detect snakes and other threats or fruits. Our long legs evolved to help us chase prey. Our lungs expanded. Our hands evolved their particular abilities in order to hold weapons. Somewhere among this series of moments, we developed the consciousness that would lead us, several generations down the road, to build cities and societies.

Our consciousness is only partial. We are conscious of the senses that we use to hunt, forage, and be social. We are not conscious of our immune system's choices, though our immune system acts in ways very similar to our other senses. None of our senses allows us full awareness of what is around us, nor, for that matter, of our decisions. We miss a range of sounds, smells, tastes, and textures obvious to other animals. We also miss most of the signals that our own eyes, noses, ears, and taste buds provide to our brains, signals that carry out on our behalf actions of which we are essentially unaware.

Oblivious to the ways our bodies depend on the species around us, we obeyed our senses. We remade the world to suit our pleasures. We excluded many species from our lives and then intentionally favored a select few, a tiny minority of Earth's diversity. Meanwhile, a suite of other species—species we now think of as pests—sneaked over our barriers and walls. They arrived, unbeckoned, in our lives. Or at least that seemed like what happened. But I have left one surprise in the story. It turns out that the species that snuck through with us into our modern lives are not a random draw from the possible life-forms on Earth. They are—these sneaky rats, roaches, and bedbugs—nearly all from the same place, though no one had noticed until recently. We had been too busy trying to kill them to pay attention to how they evolved.

In the fall of 1985, Doug Larson was hanging 600 feet above the ground. He and one of his students, Steve Spring, were looking at pine trees growing out of the cliff face, each one clinging tenuously to life. Spring was doing a rather ordinary thesis about trees that grow on rock faces, and Larson was along to mentor. Except for the precariousness of their situation, it was workaday science. The plan was to take samples of the trees to figure out how old they were. It was the student's idea. Larson would rather have been studying lichens, but he was happy to be exploring and obliging his student for a while. As Larson and his student bounced back and forth from the wall, anything might have happened, and so it was that something did.

Until his moment on the cliff, part of the Niagara Escarpment,

Larson had studied lichens for his entire professional life. They were lovely and persistent. They were also irrelevant to most of humanity, or at least they seemed that way. A case can be made for lichens, to be sure. They are a near miraculous fusion of algae and fungi, life-forms that in coming together can live in a way—on rock faces, eating air, sun, and minerals—that neither can on its own. Still, Larson found himself, again and again, having to make the case.

Then came the cliff. On the cliff, Larson and Spring sampled a few of the trees, eastern white cedars (*Thuja occidentalis*), none wider than Larson's forearm. The trees were stunted, twisted, gnarled, and otherwise beat up. It was silly, really, to study their age. They were clearly young trees, hanging on for a few years between germination and their eventual fall from the face. Or at least they seemed to be. When Larson and Spring came back to the lab, they were in for a big surprise. They looked at cores from the trees under a microscope and found not the few dozen rings they had expected, but instead hundreds of rings. The trees were hundreds of years old, an ancient forest suspended in midair. In fact, they were among the oldest trees on Earth.[5] They had resisted not just gravity, but also, somehow, time.

The consequence of discovering these trees was twofold. The age of the trees was the first piece in what would become an important and much longer story. Ancient forests would prove to reside not just on the cliffs near Larson's home, but instead on many cliffs, refuges of the past here and there around the world. Trees a thousand years old and older were found on rock faces in Canada, the United States, the United Kingdom, and France. Larson, the once-obscure scientist from Guelph, became one of the better-known Canadian biologists, a voice for interesting old trees. The other consequence though was that in looking at these trees, Larson was spurred into looking at the life on cliffs more generally. As he did, he started to view cliffs as significant, central even, to who we are as humans. A long way from any major city, he even came to develop a new theory about the emergence of our modern urban lives.

Larson's theory was a coming to terms with the similarities between his one life in an office on a campus and his other life, dangling off ledges. It was also the coming together of minds. He formed the theory along with a group of five other scientists* whose thoughts had begun to converge on his own. Together they discussed the idea excitedly, only to broaden it a little more and then, when modesty struck, rein it in and then broaden it once again. The five, with Larson as lead author, published the theory, in its nascent form, in a book on cliff ecology (the only book on cliff ecology).[6] Then, when that did not seem sufficient, they wrote an entire book about the idea, *The Urban Cliff Revolution*. In the first and most often discussed part of the book, Larson and team argue that the cities we build are like cliffs, populated with cavelike rooms and balconies. We build these clifflike environments even though they are marginal and unproductive, they argue, because through the long years of early human evolution, caves and cliff sides were our refuges from the elements and from predators. We build cities out of cement and up into the sky because they remind us of cliffs and caves. This, however, was not the only radical idea in the book.

In addition to their big idea about our fondness for caves, the team offered an explanation for the origins of the species—whether dandelions or pigeons—that live with us in our cities. They noticed that the species that make it unbeckoned into our cities tend to be the very same species that originally lived with us in caves or on cliffs. In cities all over the world, we have created a vast network of caves and cliffs into which species that evolved to live under such conditions have moved, as content and successful as they have ever been.

Cliffs occupy a tiny proportion of the earth, less than one in ten thousand acres. More land is now devoted to parking lots than to cliffsides and caves. If cliffside environments contributed randomly

*The five were Doug Larson, Uta Matthes, Peter E. Kelly, Jeremy Lundholm, and John Gerrath.

to city biotas, just one in a thousand of the species we see in cities would come from caves or cliffs. Instead, Larson has found that nearly half of the plant species in his home city grew first along cliffsides. The results for animals were similar. The list of species from cliffsides is a who's who of the life outside our windows. Dandelions, Norway rats, German cockroaches, bedbugs, plantago, peregrine falcons, rock doves (pigeons), starlings, cliff swallows, house sparrows, barn owls, earthworms (and a lovely, introduced glowworm that feeds on nothing but urban earthworms), and many more of the species that live in our cities evolved in caves and cliffs.* Some species, such as some cave crickets and firebrats, are now more common in our houses than they are in caves. These species not only happen to be the ones we have favored but they have even continued, in many cases, to live as they once lived in caves. Rock doves still nest in crevices. Their ancient predators, peregrine falcons, still stoop on them among cliff faces (albeit glass ones). They still take off "vertically and explosively," to quote Larson and colleagues, as they must for lack of runway.

This second idea of Larson's may seem like a modest biological detail, but it is not. Larson's idea about the origins of urban life-forms suggests how we might manage the life in our cities. We tend to treat life in cities as though it were simply a degraded version of nearby forests. We tend to trees and talk about "urban forestry" so often that it is now considered a field of study unto itself. The truth is, cities are something else, perhaps, as Larson suggests, a landscape that includes forest, but also far more in the way of caves and clifflike walls. If Larson and friends are right, the species we have chosen to live near us are accidentals, consequences in their success of the type of cavelike habitat we happen to have made.

*The list of species familiar to our everyday landscapes that originally lived on cliffsides or talus slopes is incredible and bears trotting out a little further—there are, on this list, tulips, geraniums, forsythias, dandelions, peonies, petunias, and then even goats, guinea pigs, and nearly all of our domestic crops, from capers and agaves to almonds, carrots, cucumbers, and even wheat.

Until now the life we appear to have gardened around us—not for food but by accident—is favored by the structure of our lives and surroundings. If right, Larson's work suggests that we need to rethink how we manage the species that live nearest to us and that we need to do so, in part, by altering the infrastructure that surrounds us. We need to favor not just the species that happen to persist, or tumble-down versions of the forests and grasslands we find farther from our cities and towns, but instead create something wilder and more interesting.

Here I will offer my own vision of what could be. I am, I admit, caught up in the revolutions of Despommier and Larson. Maybe a first course of action is to try to grow as many beneficial or potentially beneficial species as we can in cities, ideally species native to the regions in which the cities are found. They should come from cliffs and canopies (because it turns out that canopy, the dry tops of trees, is the other habitat that resembles cliffs). Imagine green walls of wild species in each and every city, even rare species, among which flit hummingbirds, butterflies, and bees. Imagine the tangled bank of wildlife rising up out of street medians. Imagine too larger green walls of life, interspersed with vertical farms. Some of these species have already made it, as harbingers of what can be done. In Hong Kong, the epiphytes that once grew exclusively on trees now grow on some downtown buildings, in great numbers (if not yet great diversity). In Mexico City, several dozen species of lichens grow on buildings and tree trunks. We could add to these species and bring with them their dependents. We could also plant fruit trees in medians, and berries could cascade off balconies. We might forage within our cities as we walk, as we once all foraged. We might remember how to gather life, both food and well-being, and if, in the process, we also gather a wider diversity of good microbes and a worm or two, so be it.

Right now, our biggest barriers remain our brains and their biases, brains that still tell us that a green pesticide-treated lawn is more healthy than one abounding with species, brains that still tell us the same things they told us when we lived in caves and when mammoths still walked along the horizon. There are logistical impediments too, some small, some big. Pollution will (as critics of Despommier's plans suggested before they realized his gardens were indoors) make some of the fruits and foods we grow in cities toxic. In Mexico City and elsewhere, it has killed lichens (in fact, lichens are used like a canary in the coal mine, an indicator of such pollution). But surely if our cities, or at least some of our cities, are too ridden with toxins to eat their fruit, the answer is to clean up our cities, not throw out our fruits. We must be able to bite into the apples that grow on street corners. We have given in to the temptations of our origins for most of our history and must now give in to something else, a vision for the future of life. If the first bite is bitter, if our first and current cities are not quite right, we should plant the seed again until what grows from our society is sweet.

If we do not succeed in preserving a rich and useful nature in and around ourselves (as our appendix attempts to do inside our bodies), the rest of nature will run us over. For all the worry about the end of nature, the persistence of life itself seems assured, at least over the next millions of years. Nature lives in hydrothermal vents and clouds. It thrives at temperatures hot enough to boil water and cold enough to freeze the marrow in our bones.

What we should worry about is the end of our nature, the links between humans and other species, links on which our very existence depends. To return to Dubos, "if we [do] not manage to create environments in which human beings, and especially children, could safely express the rich diversity of their genetic endowment," we will fail. But it is not just that our bodies miss other species and their richness. The secret that runs throughout this book, the one that I hope to have shown more than I have discussed, is that our bodies and lives only make sense in the context of other species. Only by looking at other lives do we really understand our own.

Some of the ways we look to other species to understand ourselves are so ordinary that we forget them. We do experiments on mice, guinea pigs, and rats because they are enough like us that when we understand them, we understand ourselves. But the truth is broader. Much of what we learn about ourselves comes not from lab animals, but from the wild labs of the Amazon, the Serengeti, and other places where species still mate, die, and flee of their own volition. It is in the wilderness, after all, that we see the evidence of the effects of snakes and predators on primates. It is by looking at other species that we put our appendix in context. It is in termites and other insects that the beneficial roles of microbes in guts first become most clear. It is in the ant societies that we see the most general understanding of the origins of agriculture. It is where the wild things are that we see our bodies and lives most clearly. We make sense, you and I, only in the light of our understanding of the general rules and tendencies of ecology and evolution.

How much is the understanding that we garner from the wildest places worth to us? It is hard to say precisely, but what I can say is that when we lose wild monkey species and their predators or even snakes and rare ants, we lose the most reflective mirrors with which we are able to examine ourselves. We need to maintain the kinds of wild places in which the truths about who we are are most evident. If that means rewilding the Great Plains with cheetahs so as to be able to see pronghorn once again running for a reason, so be it. Let their flight remind us that our lives, whether we notice or not, are still and will always be where the wild things are.

Acknowledgments

As humans, we tend to look straight ahead. It is a legacy of our heritage as gatherers and hunters. Looking to the periphery at our context or even looking back to where we have come from is more difficult. The view is never totally clear. Fortunately, I have been helped tremendously in my attempt to look at our context, both ecological and evolutionary, by the perspectives of others. Many individuals have helped me to see.

Harry Greene, Lynne Isbell, Geerat Vermeij, Joel Weinstock, Philip Trexler, Sarah Tishkoff, Dickson Despommier, Mark Schaller, Corey Fincher, Randy Thornhill, and Markus Rantala all generously read the chapters relating to their research and added both insight and detail. They also contributed the research that in the first place helped me to see elements of who we were and are. Maurice Pollard, Philip Trexler, Joanna Lambert, and Piotr Naskrecki spent time talking me through some of their work and also their thoughts on the relationships between the rest of life and humans. William Parker read the entire book more than once and, in doing so, became immersed in even more of the mysteries of who we are than he was before. Read Dunn, Gregor Yanega, John Godwin, Nick Haddad, Diane Dunn, John Dunn, Liz Krachey, Matthew Krachey, Michael Gavin, Jen Solomon, Philip Carter, Piotr Naskrecki, Melissa Mchale, Will Wilson, Craig Sullivan, Robert Anholt, John Vandenberg, Bob Grossfeld, Nash Turley, Marc Johnson, Cherry Crayton, and Kevin Gross all read parts or in most cases all of the book and provided valuable comments and insight.

I am thankful to Sean Menke, Benoit Guenard, Neil McCoy, Shannon Pelini, Mike Weiser, Nyeema Harris, Judith Canner, Sarah Diamond, Andrea Lucky, Jiri Hulcr, Magdalena Sorger, and other folks in my lab who know that when they knock at my door and no one answers, even when they hear typing, they should come back later. Damian Shea and Thurman Grove supported my writing as a form of an extension, another way to put, if not a person, some words out into each of the counties of North Carolina. I am especially grateful to Victoria Pryor, without whose help this book would be less rich and interesting and, in fact, might not exist at all. Elisabeth Dyssegaard and Bill Strachan edited this book into being. Thanks to my sister Jane, for spending a childhood with me out in the woods pretending to live as we all once lived, foraging for food and making our tools from what we found—at least for the few hours before our mom called us in.

And, of course, thanks to Lula and Goose for being Lula and Goose and, in doing so, reminding me just how fortunate we are to have our one wild life.

Thank you to Monica for saying yes in the back of a pickup truck near Esperanza in the Bolivian Amazon. What fun we're having with this twisting jungle ride.

Notes

1: The Origins of Humans and the Control of Nature

1. That first moment is muddied in a kind of anthropological Abbot and Costello routine about who was on first and what counts as discovery, but the very first tooth was actually found by Gen Suwa of the University of Tokyo.

2. Older sets of bones had been found, including a 6-million-year-old skull from Chad, but none of them was more than small fragments of bodies and stories. There is only so much one can extrapolate from a single disembodied head.

3. In fact, hominid fossils have been discovered from no fewer than fourteen different time intervals at or near Aramis.

4. The paper, titled *"Australopithecus ramidus*, a New Species of Early Hominid from Aramis, Ethiopia" by Tim White along with Gen Suwa at the University of Tokyo and Berhane Asfaw in the Ministry of Culture and Sports Affairs in Ethiopia appeared in the scientific journal *Nature* (371: 306–308). In this paper, what we now know as *Ardipithecus ramidus* was called *Australopithecus ramidus*, in other words, a new species of an already known genus. The novelty of the find was not yet totally clear. The name *ramidus* comes from the local Afar language in which *ramid* means "root," whether of a plant or a people.

5. Shreeve, J. June, 2010. The Evolutionary Road. *National Geographic*. http:// ngm.nationalgeographic.com/2010/07/middle-awash/shreeve-text/1.

6. In part, the question of which find represents the "earliest" human ancestor relates more to the parsing of language than to science. The earliest ancestor of humans was a single-celled microbe. What White and others mean when they discuss the "earliest human ancestor" is the first species to be more human than ape.

2: When Good Bodies Go Bad (and Why)

1. Gumpert, M. March 22, 1953. We Can Live Longer—But for What? *New York Times*.

2. Predicting and even measuring life expectancies is notoriously difficult, but at least some projections suggest declining life expectancies in Western countries in the years to come. See, for example, Olshansky, S. Jay; Passaro, Douglas J.; Hershow, Ronald C.; Layden, Jennifer; Carnes, Bruce A.; Brody, Jacob; Hayflick, Leonard; Butler, Robert N.; Allison, David B.; Ludwig, David S. 2005. A Potential Decline in Life Expectancy in the United States in the 21st Century. *New England Journal of Medicine* 352: 1138–1145.

3. Regardless of whether or not there is something to Hugot's theory, the idea that there is a specific suite of microbes that live in all our refrigerators and in refrigeration trucks and, more generally, in the connected world of cold spaces housing our food, is uncontested. They are there, dividing by the trillions each day and evolving to be a more and more prominent part of modern life, a kind of miniature Arctic, replete with food supplies. Although humans have been putting things on ice for thousands of years, the first refrigerator was not built in the United States until 1875. Rather suddenly, we shifted from eating our food relatively fresh to eating our food whenever it suited us. By 1930, millions of people in the United States owned refrigerators. Far from being "sterile," our refrigerators are full of the particular kinds of life that in fact do well in cold. When the fridge light goes off, dozens of types of bacteria, fungi, and other kinds of life actually thrive. Among the bacteria are species of *Listeria*, a potentially deadly pathogen, but also many species about which essentially nothing is known. It is because of these cold-tolerant and even cold-loving species that there is a "best by" date on your milk. We are, as we change the world, creating all sorts of new habitats into which life is shifting, whether we like it or not, forming new species and living new ways.

4. So firmly was this hypothesis believed that some patients were given frontal lobotomies. This happened horrifyingly recently. A study in 1956 reported the results of lobotomies on six patients who were suffering from both psychological problems and, independently, Crohn's disease. The intestinal problems of three patients got better (chance would predict that three would get better and three worse). Two patients died. The experiment was viewed as a success.

5. E-mail from J.V. Weinstock, May 18, 2009.

6. Vaughan, T. A. 1986. *Mammalogy*. 3rd ed. New York: Harcourt Brace Jovanovich College Publishers.

7. Bucky and other details of the Byerses' lives and study of pronghorn come

from e-mails from John Byers and his lovely book *Built for Speed, a Year in the Life of Pronghorn*. 2003. Cambridge, Mass.: Harvard University Press.

8. Here he is quoting Willa Cather, who wrote, "Elsewhere the sky is the roof of the world; but here the earth was the floor of the sky," in *Death Comes for the Archbishop*.

9. For example, see Lindstedt, S. L.; Hokanson, J. F.; Wells, D. J.; Swain, S. D.; Hoppeler, H.; and Navarro, V. 1991. Running Energetics in the Pronghorn Antelope. *Nature* 353: 748–750.

10. Mammal biologists love to catalogue deaths of their study organisms. For baby pronghorn, see Beale, D. M., and Smith, A. D. 1973. Mortality of Pronghorn Antelope Fawns in Western Utah. *Journal of Wildlife Management* 37: 343–352, in which about one in four fawns were eaten by bobcats.

11. This "cheetah" is actually unrelated to the African cheetah. Research by Barnett and colleagues shows that the American cheetah is most closely related to the cougar. The American cheetah evolved traits similar to those of the African cheetah—including elongated limbs, enlarged nasal passages, and claws that do not retract—independently. These African and American cheetahs are an example of convergent evolution, wherein two species evolve similar traits in response to similar conditions. The conditions in this case were the expansion of grasslands and with them, the evolution of species like the pronghorn or its convergent African analogue, the antelope, both of which avoid predators by fleeing at high speeds. Barnett, R.; Phillips, M. J.; Martin, L. D., Harington, R.; Leonard, J. A.; and Cooper, A. 2005. Evolution of the Extinct Sabretooths and the American Cheetah-like Cat. *Current Biology* 15: R589–R590. doi:10.1016/j.cub.2005.07.052. http://linkinghub.elsevier.com/retrieve/pii/S0960982205008365. Retrieved 2007-06-04.

12. A recent study by J. J. Dennehey has shown that low-ranking females tend to be forced to the margins of the herd to forage on their own. They are shunned and solitary. Historically, such individuals would have been the most likely to have been eaten, the bad luck of being low on the totem pole. But today, with little threat of predation, those low-ranked individuals actually obtain better quality food than those at the middle of the hierarchy who are safe from nonexistent predation but have to fight for food alongside the very highest ranked individuals. It is conceivable that these realities will, in the long term, affect the evolution of pronghorn social structure and with it, their speed. 2001. Influence of Social Dominance Rank on Diet Quality of Pronghorn Females. *Behavioral Ecology* 12:177–181.

13. And slightly adulterate . . . Martin was referring specifically to North America, but the sentiment seems to generalize.

14. More about the anachronistic fruits in particular is detailed in Connie Barlow's book *The Ghosts of Evolution, Nonsensical Fruit, Missing Partners, and other Ecological Anachronisms.* (New York: Basic Books, 2000).

3: The Pronghorn Principle and What Our Guts Flee

1. The taboos of conservative science are not necessarily bad. Science is slow to accept new ideas and initially regards radical new ideas as taboo, but with good reason. For every brilliant idea that seems ridiculous there are a thousand ridiculous ideas that look and are ridiculous. What separates science from nonscience is in part the stringent filters that winnow away the ridiculous and in the process occasionally also toss a revolutionarily good idea or two.
2. Hunter was writing about mammoths and mastodons, but his sentiment seems more general.
3. Hansen, D. M.; Kaiser, C. N.; and Müller, C. B. 2008. Seed Dispersal and Establishment of Endangered Plants on Oceanic Islands: The Janzen-Connell Model, and the Use of Ecological Analogues. *PLoS ONE* E222 http://www.plosone.org/article/info%3Adoi%2F10.1371%2Fjournal.pone.0002111.
4. Summers, R. W.; Elliot, D. E.; Urban, J. F.; Thompson, R.; and Weinstock, J. V. 2005. *Trichuris suis* Therapy in Crohn's Disease. *Gut* 54: 87–90.
5. Five of those patients had gone into remission and the sixth improved. Hopeful, but very preliminary results were presented at the American Gastroenterological Association meetings in May 1999. According to a statement made by Weinstock in an article in the *New York Times*, the patients were begging to get more worms after the trial.
6. Because of doctor-patient confidentiality, there is no way to know whether some, all, or none of these patients are still doing better.
7. Saunders, K. A.; Raine, T.; Cooke, A.; and Lawrence, C. E. 2007. Inhibition of Autoimmune Type 1 Diabetes by Gastrointestinal Helminth Infection. *Infection and Immunity* 75: 397–407.

4: The Dirty Realities of What to Do When You Are Sick and Missing Your Worms

1. These symptoms were apparently the result of magnesium depletion, but why the magnesium depletion occurred in the first place is unclear.
2. Incan skull surgeries were, in the later years in which they were practiced, actually quite successful. But this was only after a few hundred years of trial and error, the error in this case leading to a very bad day.

5: Several Things the Gut Knows and the Brain Ignores

1. Such as the fox and coyote and other animals that biologists tend to call "meso-predators" where meso means intermediate. When meso-predators do better in the absence of bigger, badder predators it is referred to as meso-predator release.

2. Survey carried out in 2000 by the European Federation of Animal Health (FEDESA).

3. In 1928, Howard Walter Florey and Ernst Borish Chain shared the prize with Fleming. Fleming, who was a brilliant bacteriologist but kept a sloppy lab, had left a stack of petri dishes growing *Staphylococcus* bacteria in a pile in the corner while he went on vacation with his family. When he came back from a trip, he noticed that some of the dishes had grown over with fungi. The fungi had apparently killed some of the *Staphylococcus* in the petri dishes into which it had invaded. Fleming began to eagerly study this fungus, which he referred to as "fungus juice" (later to be named penicillin). In doing so, he would learn more about it, including that it was difficult to isolate, as were its active components. Fleming was able to show that penicillin could kill many kinds of bacteria, but he had less success making it useful as a treatment. He needed help from a chemist to isolate the active component in penicillin, but no one would help and so, in essence, he worked on the project for twenty years before abandoning it. It was then that Florey and Chain, unaware that Fleming was even still alive, took up the challenge of isolating penicillin. To this strange story, we owe penicillin and millions of human lives.

4. The four treatments were "(no antibiotics), streptomycin (0.5 g/250 ml drinking water), streptomycin-bacitracin (0.5 g of each/250 ml drinking water), and vancomycin (0.125 g)-neomycin (0.25 g)-metronidazole (0.25 g)-ampicillin (0.25 g, combined in 250 ml water)."

5. Some microbiologists now argue that the distinction between "bad" and "good" bacteria is a false one. Whether they are bad or good depends more on where they are than who they are. *E. coli* in the gut at low densities is all right, but at high densities or inside the body cavity it can be deadly.

6. Thone, F. 1937. Germ-free Guinea Pigs. *Science News Letter* 31: 186–188.

7. This is what Reyniers believed. The truth was more complicated, as it often is. In the late 1800s, two Germans, George Nuttal and H. Thierfelder tried an approach very similar to the one Reyniers was about to undertake. Simultaneous with Reyniers's work (and also apparently unknown to him), a group in Sweden led by Bengt Gustafsson and his professor, Gusta Glimstedt, was attempting to establish germ-free rats. The history of these other attempts at

germ-free life is well chronicled by Philip B. Carter and Henry L. Forster in their chapter Gnotobiotics in the now classic (and fascinating, I promise) book, *The Laboratory Rat* by M. A. Suckow, S. Weisbroth, and C. L. Franklin, the 2nd edition of which was published in 2007 by CRC Press in Boca Raton, Fla.

8. From an interview with Philip Trexler on June 11, 2010, when he was ninety-eight years old. Trexler was to become a man behind the scenes at the Lund Institute, where he initially helped Reyniers and then eventually took over much of the work of innovating new technologies. Trexler would be key, in particular, in developing technologies to be used during human surgeries to create miniature germ-free environments on humans, though only for a moment.

9. http://www.time.com/time/magazine/article/0,9171,883334,00.html.

10. By 1937, he was to be the sole faculty member in charge of a 5,000-square-foot facility dedicated solely to the study of germ-free animals. It was a building of germ-free life. In it, by 1950 he was in charge of even more space and people as well as an individual chamber large enough to house 1,000 animals including chickens, guinea pigs, and rats, but by then also monkeys. Scientists had to dive through a vat of antibiotics in order to get into the large chambers, as though passing into a brave new world. Germ-Free Animal Colony Begun in Notre Dame Tank. *New York Times*, June 22, 1950.

11. Experiments done by Dr. J. R. Blayney in Reyniers's group would be the first to show that bacteria were responsible for tooth decay.

12. Gordon, H. A., and L. Pesti. 1971. The Gnotobiotic Animal as a Tool in the Study of Host Microbial Relationships. *Bacteriological Reviews* 35: 390–429.

13. Though he would commit some follies because of hubris. It was one of those follies that would get him fired from Lobund, fired for what the chancellor at the time referred to as "financial flim flam." It was also that hubris that caused Reyniers to continue to favor the production, sales, and advocacy of metal germ-free chambers that cost thousands of dollars even when Philip Trexler had figured out how to produce plastic chambers at a cost of a few hundred dollars. The metal chambers made Reyniers's family machine-shop money. The plastic chambers would not.

14. Rukhmi, V.; Bhat, C.; and Deshmukh, T. 2003. A Study of Vitamin K Status in Children on Prolonged Antibiotic Therapy. *Indian Pediatrics* 40:36–40.

15. For a review of this exciting new area of research, see F. Bäckhed. 2009. Addressing the Gut Microbiome and Implications for Obesity. *International Dairy Journal* 20: 259–261. Yes, this means there is, in fact, a dairy journal.

16. There are actually, I believe, two times in the history of biology when science

threatened to dissolve as a consequence of individuals speaking many languages. In the first case, one I write about in *Every Living Thing*, individual scientists were naming plants and animals with different names in different countries (and even within the same country). There was no common language for naming life, and for a while it wasn't clear that one would emerge. The second occasion is, I believe, the situation in which we currently find ourselves, when different fields of science in general, but biology in particular, are so mutually unintelligible that scientists in one field tend to turn to popular (rather than scholarly) accounts of other fields to understand them.

17. Individual scientific meetings can sometimes now be so large that they do not fit into any individual conference centers. The big U.S. meeting for neurobiologists now includes more than 60,000 scientists, nearly all of them focused on some topics sufficiently narrow that at most a few hundred other scientists at the meeting find them interesting.

6: I Need My Appendix (and So Do My Bacteria)

1. Even articles that attribute some role to the appendix typically do so in vague terms. A 2001 article in *Scientific American* concludes that "a growing quantity of evidence indicates that the appendix does in fact have a significant function as a part of the body's immune system" but then does not go on to even speculate what that "function" might be.

2. Some have suggested that the appendix cannot go away because a small appendix is more dangerous and likely to burst than a long one, but the presence of small appendices in many, many species makes clear that this is not necessarily true.

3. Animals have made the evolutionary transition to cave life many tens or even hundreds of times, and each time their eyes have become reduced, their bodies pale. All that is unnecessary and costly has been lost, leaving these creatures like pale ghosts of evolution's fastidiousness bumping around in the eternal night just under the ground.

4. William Parker and colleagues have made a recent attempt to reconstruct the evolution of appendices on the mammal evolutionary tree. If you want to take your own stab at trying to discern what the species with appendices share, I direct you to this new paper in which the species are listed. They almost inevitably share something. But what? Two of the species feed on bark, so perhaps they are unique. If one throws them out, then most of the species live in habitats in which diseases might be expected to be a greater than normal threat (e.g., in the dirt or in societies), but there are other organisms in

the same habitats with no sign of an appendix. Mysteries remain. Go ahead and have a look yourself. Smith, H. F.; Fisher, R. E.; Everett, A. D.; Thomas, R.; Bollinger, R.; and Parker W. 2009. Comparative Anatomy and Phylogenetic Distribution of the Mammalian Cecal Appendix. *Journal of Evolutionary Biology* 22: 1984–1999.

5. Nor is Picasso's story really so simple as that of the youthful flame. He painted well into his nineties, as did Chagall. Monet painted into his eighties. Similar examples exist for music (Richard Straus), film (John Huston), and, of course, literature (Saul Bellow). For a lovely discussion of the fruits of age see the May 21, 2005, *New York Times* article by Alan Riding; http://www.nytimes.com/2005/05/21/arts/design/21mati.html?pagewanted=1&_r=1.

6. This sorting of "us" and "them" is identical to what worker ants or termites do in identifying invaders. Like our antibodies, they make such assessments chemically, in this case using sensors on their antennae and elsewhere on their bodies to cue in on what is foreign and what is not. Foreign species are then, more often than not, mercilessly attacked.

7. For a delightful (if slightly grumbling) discussion of one aspect of how fictions get recorded as fact in science, see Slobodkin, L. 2001. The Good, the Bad, and the Reified. *Evolutionary Ecology Research* 3: 1–13. Larry Slobodkin was my adviser's adviser and a very clever scientist and rhetorician. His rhetoric was so good, in fact, that I think it predisposed his occasional scientific hand waving to going from idea to fact well before being tested. Slobodkin, for example, is the one who suggested that 10 percent of energy goes from primary producers (e.g., grass) to herbivores (cows) and then 10 percent of energy goes from herbivores to predators (e.g., mountain lions). That number, 10 percent, became magic and true, and is reprinted in virtually every textbook of basic biology. It is reprinted even though it is not remotely accurate. Ecological systems vary widely in the percentage of energy that goes from one level in the food chain to the next. And even if we take the average of different systems or species, it does not approach 10 percent. In his last years, Slobodkin railed against this and others of his youthful assertions that had become encodified in science, but to no avail. They remain in textbooks, fiction even though we know otherwise, a complicated legacy of a brilliant man.

8. Sonnenburg, J. L.; Angenent, L. T.; and Gordon, J. I. 2004. Getting a Grip on Things: How Do Communities of Bacterial Symbionts Become Established in Our Intestine? *Nature Immunology* 5: 569–573.

9. Palestrant, D.; Holzknecht, Z. E.; Collins, B. H.; Parker, W.; and Miller, S.

E. 2004. Microbial Biofilms in the Gut: Visualization by Electron Microscopy and by Acridine Orange Staining. *Ultrastructural Pathology* 28: 23–27.

10. The standard reference on immunology—*Atlas of Immunology*, by Julius M. Cruse, and Lewis, Robert E. Boca Raton, Fla.: CRC Press, 2004. — does not even list the word "mutualism" in the index. The possibility of cooperation has been expunged from the medical lexicon.

7: When Cows and Grass Domesticated Humans

1. Interestingly, although it seems clear that most groups knew many tens and often hundreds of species of plants and animals and their uses, the numbers of known medicinal plants may have been low until agriculture arose. It was with agriculture that the need for medicine to treat disease arose and so too the knowledge.

2. Denevan, W. 1992. The Aboriginal Population of Amazonia. Pages 205–234 in Denevan, W. M., ed. *The Native Population of the Americas in 1492*. Madison: University of Wisconsin Press.

3. The Amazon and its people had long and falsely been assumed "pristine," but most Amazonian soil bears the marks of charcoal from human burning. In some places, when roads are plowed through the hills, potshards are so dense that they pour out like candy from a piñata.

4. A time budget for Hadza women, for example, includes just forty-two hours a week of work and that work includes everything—gathering food, preparing food, caring for children, and tending to the repair and construction of homes. More to the point, the Hadza seem to work *more* than did and do other hunter-gatherers.

5. Clutton-Brock's book is a lovely treatise on the biology of those species we depend on disproportionately. It is well worth a read even for someone just casually interested in those few species whose survival we have often depended on. Clutton-Brock, J. 1999. *A Natural History of Domesticated Animals*. Cambridge, U.K.: Cambridge University Press.

6. The literature on lactase persistence is growing rapidly. For a clear-headed and comprehensive view of the literature and the more general story of human genetic diversity (particularly in Africa), I recommend Scheinfeldt, L. B.; Soi, S.; and Tishkoff, S. A. 2010. Working Toward a Synthesis of Archaeological, Linguistic, and Genetic Data for Inferring African Population History. *Proceedings of the National Academy of Sciences* 107: 8931–8938.

8: So Who Cares If Your Ancestors Sucked Milk from Aurochsen?

1. With 280,000 deaths a year attributed to obesity in the United States.
2. Hammer, K., and Khoshbakht, K. 2005. Towards a "Red List" for Crop Plant Species. *Genetic Resources and Crop Evolution* 52: 249-265.
3. One piece of evidence in support of this idea is that other primates almost universally have few copies.
4. Zimmet, P.; Alberti, K. G. M. M.; and Shaw, J. 2001. Global and Societal Implications of the Diabetes Epidemic Lifestyle, Overly Rich Nutrition and Obesity. *Nature* 414: 782–787.
5. Yu, C. H. Y., and Zinman, B. 2007. Type 2 Diabetes and Impaired Glucose Tolerance in Aboriginal Populations: A Global Perspective. *Diabetes Research and Clinical Practice* 78: 159-170.
6. See, for example, Scheinfeldt, L. B.; Soi, S.; and Tishkoff, S. A. 2010. Working Toward a Synthesis of Archaeological, Linguistic, and Genetic Data for Inferring African Population History. *Proceedings of the National Academy of Sciences* 107: 8931–8938.
7. Because race is construed differently in different places, the ways in which we mistakenly use race in medicine differ from country to country. In truth, the differences among peoples relate to either genetics (and hence history) or culture, neither of which is encapsulated by race, particularly as race is often measured in a medical setting by a doctor or nurse's checkbox assessment. Braun, L.; Fausto-Sterling, A.; Fullwiley, D.; Hammonds, E. M.; Nelson, A.; Quivers, W.; Reverby, S. M.; and Shields A. E. 2007. Racial Categories in Medical Practice: How Useful Are They? *PLoS Med* 4(9): e271. doi:10.1371/journal.

9: We Were Hunted, Which Is Why All of Us Are Afraid Some of the Time and Some of Us Are Afraid All of the Time

1. I gave Bakhul her name. Her real name was lost to history.
2. Fitzsimons, F. W. 1919. *The Natural History of South Africa*. New York: Longmans, Green and Co.
3. Immortalized in several books as well as in two recent films, the most recent of which was the very successful 1996 *The Ghost and the Darkness*.
4. Tongue, M. H. 1909. *Bushman Paintings*. Oxford: Clarendon Press.
5. There is an old joke in which Johnny and Pete are out in the woods hiking in boots. A grizzly bear starts chasing them. Johnny stops to change into his tennis shoes and Pete yells, "What are you doing, Johnny? You can't outrun a bear," to which Johnny responds, "I don't have to outrun the bear. I just

have to outrun you." This joke cuts to the core of who we are: just another animal who has spent generations figuring out strategies to avoid being the buddy who gets eaten by the bear. Our fear module evolved to allow us to be the one who gets away. Our neocortex, the conscious man's front brain, evolved to give us the creativity to invent and put on shoes.

6. McDougal, C. 1991. Man-eaters. In *Great Cats: Majestic Creatures of the Wild.* John Seidensticker and Susan Lumpkin, consulting editors. Emmaus, Pa.: Rodale Press.

7. More recent, if gruesome, studies of baboons fed experimentally to leopards confirm that modern leopards consistently leave the heads of their primate prey and regurgitate the other bones. The fingers tend to come back up intact, as was also the case with the bones discovered in the South African caves. Carlson, K. J., and Pickering, T. R. 2007. Intrinsic Qualities of Primate Bones as Predictors of Skeletal Element Representation in Modern and Fossil Carnivore Feeding Assemblages. *Journal of Human Evolution* 44: 431–450.

8. This latter study included a drawing with the caption, "Reconstruction of a leopard dragging an . . . ape-manchild. It is suggested that the damage-marks found on the child's skull could have been caused by the lower canines of the leopard when holding the child's head in the position shown." This caption does not seem to actually require a picture. Brian, C. K. 1969. *South African Archaeological Bulletin* 24: 170–171.

9. Among them, *Agriotherium* (a giant, dog-faced bear), *Chasmaporthetes* (a fast-running, hyena-like carnivore), *Machairodus* (a saber-toothed cat), *Dinofelis* (another saber-toothed cat), *Homotherium* (yet another saber-toothed cat), *Pachycrocuta* (a group of hyenas, including the giant hyena), and *Megantereon* (a cat built like a modern jaguar).

10. Jenny, D., and Zuberbuhler, K. 2005. Hunting Behaviour in West African Forest Leopards. *African Journal of Ecology* 43: 197–200.

11. Isbell, L. A. 1994. Predation on Primates: Ecological Patterns and Evolutionary Consequences. *Evolutionary Anthropology* 3: 61–71. It is perhaps worth noting that some gorillas do occasionally make night nests in trees, but they are the young gorillas, who are still small enough to climb well and/ or to be eaten easily.

12. Alrod, P. L.; Nash, L. T.; Fritz, J.; and Bowen, J. A. 1992. Effects of Management Practices on the Timing of Captive Chimpanzee Births. *Zoo Biology* 11: 253–260.

13. Interestingly, not all domesticated animals have become so numbed. Horses remain twitchy, their tendency to flee ever ready, unbroken. In part, these

differences reflect the traits we have favored in different domesticated animals. In horses (and other animals used in transportation, such as camels and donkeys), we wanted speed and power. In cows, sheep, and pigs, we wanted more simply milk and meat.

14. Domestication Effects on Foraging Strategy, Social Behaviour and Different Fear Responses: A Comparison between the Red Junglefowl (*Gallus gallus*) and a Modern Layer Strain. *Applied Animal Behaviour Science* 74: 1–14.

10: From Flight to Fight

1. Mediterranean tortoises, for example, take around ten years to reach reproductive age. Even limpets and drills take several years. These animals, though once abundant, were slow to recover from being hunted, and so are now abundant almost nowhere in the world. Stiner, M. C.; Munro, N. D.; and Surovell, T. A. 2000. The Tortoise and the Hare: Small Game Use, the Broad Spectrum Revolution, and Paleolithic Demography. *Current Anthropology* 41:39–73.

2. Young, R. W. 2003. Evolution of the Human Hand: The Role of Throwing and Clubbing. *Journal of Anatomy* 202: 165–174.

3. Corbett went on to kill many "man-eating" tigers, leopards, and lions, but he was also an adamant advocate for the conservation of big cats. To Corbett there seemed in the modern story to be an uneasy truce between big cats and humans, a truce violated by the cats only when they were sick and old but violated by humans at every opportunity.

11: Vermeij's Law of Evolutionary Consequences and How Snakes Made the World

1. Wayne, R. K.; Benveniste, R. E.; Janczewski, D. N.; and O'Brien, S. J. 1989. Molecular and Biochemical Evolution of the Carnivora. In Gittleman, J. L., ed., *Carnivore Behavior, Ecology, and Evolution.* Ithaca, N.Y.: Cornell University Press, pp. 465–494.

2. Isbell, L. 1994. Predation on Primates: Ecological Patterns and Evolutionary Consequences. *Evolutionary Anthropology.* 3: 61–71.

3. Andersen, P. R.; Barbacid, M.; and Tronick, S. R. 1979. Evolutionary Relatedness of Viper and Primate Endogenous Viruses. *Science* 204: 318–321.

4. See Greene's beautiful treatise on snakes and their sublimity: 1997. *Snakes, the Evolution of Mystery in Nature.* Berkeley: University of California Press.

5. The study of the "persistent effects of the idiosyncratic distribution of life" is

the bread and butter of my own field, biogeography, where *bio* means life, *geo* means Earth, and *graphy* relates to the story of, and so the field might be described as the story of Earth as conveyed by life. It is one such story, that of humans, that I am revealing here, our story as told by our interactions, across space and time, with the rest of life.

6. Vermeij, G. J. 1977. Patterns in Crab Claw Size: The Geography of Crushing. *Systematic Zoology* 26: 138–151.

7. Notably, if you want to understand a little about what shells were like before crabs with big claws evolved in the ocean, you can look to ponds. In ponds, crushing predators are rare, so one can still find snails with unprotected, simply coiled, shells with wide openings. There they live as if in some more innocent past.

8. Vermeij first elaborated his idea in 1982 in the paper Unsuccessful Predation and Evolution in the *American Naturalist* (120: 701–720). Vermeij never refers to his idea as a law. That is my doing.

9. In fact, hornbills (a large forest bird) can also recognize Diana monkey calls. When Diana monkeys scream, "big cat," the hornbills do not respond (hornbills are never eaten by leopards). But when Diana monkeys scream, "big bird," the hornbills begin to scream as well and look for the bird.

10. Isbell, L. A. 2009. *The Fruit, the Tree and the Serpent: Why We See So Well.* Cambridge, Mass.: Harvard University Press.

11. Isbell, for her part, sees fruit consumption and snakes as linked. Once vision began to improve because of the need to detect snakes, she thinks that fruit (which could then also more easily be detected) provided the energy necessary for bigger and bigger brains.

12. Nor are the effects of snakes on our biology due to venomous snakes alone. Harry Greene, in an e-mail (June 15, 2010) informed me that he was nearing publication of a paper showing that the indigenous Agta people of the Philippines were long plagued by pythons. Of 120 people whose lives were studied, 26 percent of adult males had been attacked by reticulated pythons. Six of those attacks were fatal.

12: Choosing Who Lives

1. See Wu, S. V; Rozengurt, N.; Yang, M.; Young, S. H.; Sinnett-Smith, J.; and Rozengurt, E. 2002. Expression of Bitter Taste Receptors of the T2R Family in the Gastrointestinal Tract and Enteroendocrine STC-1 cells. *Proceedings of the National Academy of Sciences* 99: 2392–2397. Other species have an even

greater diversity of taste bud locations. Sturgeon have taste buds on the outsides of their lips and so, unlike us, are able to taste food before bothering to put it into their mouths. Catfish have taste buds all over their bodies. To them, the whole world is a lunch.

2. Dean, W. R. J.; Siefried, W. R.; and MacDonald, A. W. 1990. The Fallacy, Fact, and Fate of Guiding Behavior in the Greater Honeyguide. *Conservation Biology* 4: 99–101.

3. Leff, B.; Ramankutty, N.; and Foley, J. A. 2004 Geographic Distribution of Major Crops across the World. *Global Biogeochemical Cycles* 18: GB1009, doi:10.1029/2003GB002108.

4. Interestingly, this struggle can vary slightly from person to person as a function of our particular histories. Some individuals can taste the chemical compound PTC (it registers as bitter), whereas others cannot. This variation is the result of genetic differences in a series of known genes. This difference may be adaptive. Individuals who can taste PTC are also better able at tasting (and spitting out) plants that contain bitter toxins and so may have done better where toxic plants were diverse. At the same time, in our modern setting this gene has fewer advantages and some disadvantages. Individuals who can taste PTC are much less likely to enjoy some vegetables, such as broccoli, which have an excess of defensive compounds.

5. Though it is interesting to note that humans (and perhaps some other mammals) have the ability to learn to like some bitter and sour tastes, as is the case with coffee. Just how such preferences are learned is not yet clear.

6. This is not to say that thirst or hunger in our modern environments are entirely rational either. To the extent that our hunger has an equilibrium, it tends to be about 3,000 calories in men and 2,000 in women. With that many calories, we tend to be, on average, full. But the fact that our bodies are full after 2,000 or 3,000 calories evolved in a context when we searched for and chased down our food. We do not move as much anymore, but our hunger system remains unchanged, irrational but the same. Interestingly, exercise is a way of reenacting our former activity levels, both to use our muscles in ways similar to those in which they evolved and to burn the number of calories for which our bodies ask. Some academics have gone so far as to argue it is to remedy the discrepancy between who we were and who we are that we began to exercise in the first place.

7. DeLoache, J. S., and LoBlue, V. 2009. The Narrow Fellow in the Grass: Human Infants Associate Snakes. *Developmental Science* 12: 201–207 DOI: 10.1111/j.1467-7687.2008.00753.

8. Morris, J. S.; Öhman, A.; and Dolan, R. J. 1999. A Subcortical Pathway to

the Right Amygdala Mediating "Unseen" Fear. *Proceedings of the National Academy of Sciences* 96: 1680–1685.

13: How Lice and Ticks (and Their Pathogens) Made Us Naked and Gave Us Skin Cancer

1. Weiss, R. A. 2009. Apes, Lice and Prehistory. *Journal of Biology* 8:20.
2. Kushlan, J. A. 1980. The Evolution of Hairlessness in Man. *American Naturalist* 116: 727–729.
3. Malaria means bad (mala) air (aria), though given the dependence of malaria-transmitting mosquitoes on pools of stagnant water, *malaqua* might have been more apt. *P. falciparum* is, in fact, just one of several malarias.
4. This story is even more complicated and wonderful than I have space to discuss. For additional information please read . . . DOI: 10.1126/science .1063292. Luzzatto, L., and Notaro, R. 2001. Protecting Against Bad Air. *Science* 293: 442–443.

14: How the Pathogens That Made Us Naked Also Made Us Xenophobic, Collectivist, and Disgusted

1. Thornhill, R., and Alcock, J. 1983. *The Evolution of Insect Mating Systems.* Cambridge, Mass.: Harvard University Press.
2. Thornhill, R., and Palmer, C. T. 2000. *A Natural History of Rape: Biological Bases of Sexual Coercion.* Cambridge, Mass.: MIT Press.
3. The term "behavioral immune system" would not be coined until later, by Mark Schaller, but the idea was already present, albeit vaguely.
4. For a nice, if somewhat dated, review of the responses of animals to parasites and diseases see Hart, B. L. 1992. Behavioral Adaptations to Parasites: An Ethological Approach. *Journal of Parasitology* 78: 256–265.
5. Fincher, C. L.; Thornhill, R.; Murray, D. R.; and Schaller, M. 2008. Pathogen Prevalence Predicts Human Cross-cultural Variability in Individualism/Collectivism. *Proceedings of the Royal Society B: Biological Sciences* 275: 1279–1285.
6. Schaller, M., and Murray, D. 2008. Pathogens, Personality and Culture: Disease Prevalence Predicts Worldwide Variability in Sociosexuality, Extraversion, and Openness to Experience. *Journal of Personality and Social Psychology* 95: 212–221.
7. Schaller, M.; Miller, G. E.; Gervais, W. M.; Yager, S.; and Chen, E. 2010. Mere Visual Perception of Other People's Disease Symptoms Facilitates a More Aggressive Immune Response. *Psychological Science* 21: 649–652.

8. Duncan, L. A., and Schaller, M. 2009. Prejudicial Attitudes toward Older Adults May Be Exaggerated When People Feel Vulnerable to Infectious Disease: Evidence and Implications. *Analyses of Social Issues and Public Policy* 9: 97–115.

15: The Reluctant Revolutionary of Hope

1. Unlike the worms being used to treat diseases associated with our immune systems, *Trichinella* did not evolve to live in humans. It is a pig worm, and we only encounter it when we eat pigs. Once inside our body, it is not sure what to do. Lost, it gets up to trouble, which is what makes us sick, though we tend to fare better than it does. Inside a human, it always dies, whereas with *Trichinella* inside us we only sometimes do.
2. http://www.downtownexpress.com/de_133/greenroofsaregrowig.html.
3. Ryerson University. 2009. Report on the Environmental Benefits and Costs of Green Roof Technology for the City of Toronto. http://www.toronto.ca/greenroofs/findings.htm.
4. Skyfarming. *New York Magazine* http://nymag.com/newsfeatures/30020/#ixz zoaUH4bkTj.
5. Larson, D. W.; Matthes, U.; Gerrath, J. A.; Larson, N. W. K.; Gerrath, J. M.; Nekola, J. C.; Walker, G. L.; Porembski, S.; and Charlton, A. 2000. Evidence for the Wide-spread Occurrence of Ancient Forests on Cliffs. *Journal of Biogeography* 27: 319–331.
6. Larson, D. W.; Matthes, U.; and Kelley, P. E. 2000. *Cliff Ecology*. Cambridge, U.K.: Cambridge University Press.

Index

bitter taste, 183, 189–90
blindsight, 193–94
blood pressure, 189
Bolivia, 233–36
Bollinger, Randal, 97–99, 103, 104, 105
Bonaparte, Napoleon, 212
Booker, Corey, 250–51
brain:
  evolution of, 175–76
  and fear response, 153, 195, 250
Brazil, snakebites in, 178
breast milk, 80
bubble boy, 76, 84
buffalo, hunting, 158
Byers, John, 24–28, 30, 31, 39, 67
Byers, Karen, 24–27, 30, 31

calories:
  from microbes, 83
  sources of, 130
  and survival, 132
camels, extinct, 28, 34
cancer, deaths from, 149
CARD15 gene, 19
carnivores, teeth of, 30
carrion, 191, 210
Cassia grandis tree, 29
cats:
  saber-toothed, 27, 61
  taste buds of, 183
  and *Toxoplasmosis gondii*, 149-50n
cave fish, 94–95, 176
cave lions, giant (*Panthera atrox*), 27
caves, 210–11, 255, 256
Cavineños, 139
cedars, eastern white (*Thuja occidentalis*), 254
cellulose, 77, 79
chance:
  and germ-free animals, 79, 81
  and inheritance, 136
cheese, 126

cheetahs:
  American (*Miracinonyx trumani*), 27–28, 61
  giant, 27
cherimoya, 29
chickens, 154n, 209
chimpanzees:
  births of, 151
  eating monkeys, 174
  and human evolution, 7
China, sparrows in, 158-59n
chlamydia, 208
choices, 196–99
cholera, 98–99, 104
circulatory system, evolution of, 188–89
cities:
  future of, 234
  green rooftops in, 240–42, 251
  growing food in, 244, 246
  of insects, 245
  pest species in, 248
  pollution in, 258
  public gardens in, 244
  restoring nature in, 249
  species of, 256
  structure of, 233
  urban cliffs, 255–57
  vertical gardens in, 245–46, 248, 250–51, 257
civilization, and agriculture, 114–16
cliffs, life on, 254–57
Clutton-Brock, Juliet, 120
cockroaches, 248
collectivism, 222, 223, 226–27, 228
colon, biofilm in, 103, 104
Columbus, Christopher, 219
Colwell, Rob, 177
commensals, 68
communication, failure of, 85–86
consciousness, 252–53
Copenhagen, Denmark, 19

Earth, impact of humans on,
12–13
E. coli, 226
ecology, human connections to,
x, xi
ectoparasites, 205–16, 220, 223
Ehrlich, Paul, *The Population Bomb,*
x
elephants, 206
endoparasites, 208
environment, created, 233–34
enzymes, 79–80
Eosimias, 178
evolution:
  and aching, xii–xiii
  and agriculture, 138
  of brains, 175–76
  of circulatory system, 188–89
  in ecological contexts, 136
  and fear module, 147–48
  and genetics, 121, 127, 128
  human history in, x, 252–53
  and hunting, 157–58
  and interspecific interactions, 30
  and language, 175
  natural selection, 26, 28, 79
  of plants, 40
  and universals, 181–82
  and Vermeij's law, 172–73, 174
  of vision, 175–76, 179
excrement, smell of, 190–91
extinction, 11, 12–13, 28
  of cultures, 138
  and domestication, 124–25
  and interspecific interactions, 30
eyelashes, 209
eyes, quality of, 164–65, 179

farming, 12–13
fat:
  desire for, 189
  storage of, 134–35
  usefulness of, 134
Faulkner, Jason, 221, 222, 224

fava beans, 215
favism, 216
fear:
  misplaced, 161–63
  phobias, 162
fear module, 145–46, 147–48,
  152–54, 161, 195, 250
fiber, 80
fight-or-flight response, 145–46,
  152–53, 154, 162, 199
figs, 10n
Fincher, Corey, 218–22, 223–24,
  226, 227, 228
fire, 12, 196, 198
fire ants (*Solenopsis invicta*), 82
fleas, 209–10, 211, 213
Fleming, Alexander, 62
flies, 205, 206, 213, 214
food:
  digestion of, 78, 79, 80, 81
  and germ-free animals, 79–80
  hunger for, 190
  production, *see* agriculture
food pyramid, 132
fossils:
  discovery of, 3–10
  predator evidence in, 148–49
  in sediment layers, 7, 10
foxes, and domestication, 154
fruit, evolution of, 29
fruit flies, and human genetics, 7
Fulani people, 126
fungi:
  (*Escovopsis*), and ants, 89
  in human bodies, 82
  and lichens, 254
fur mites, 212

G6PD gene, 215
Ganesh, 207
genetic code, 123
genetics:
  and diversity, 137–38
  and evolution, 121, 127, 128

Mao Tse-tung, 158–59n
Maple, Maura, 177
Martin, Paul S., 29
Masai:
  as cattle people, 139
  drinking milk, 125–26
mastodons:
  extinct, 28
  hunting, 61
Matisse, Henri, 97
Matthes, Uta, 255n
Mbuti pygmies, 151
meat, eating, 133
melanin, 204, 213–14
mermaids, 205
metabolism, 131, 135
Mexico City, lichens in, 257, 258
mice, experiments on, 36, 39,
    65–66, 81–82
microbes:
  calories from, 83
  and digestion, 78, 79, 80, 81, 83
  as mutualists, 85, 106
  of our ancestors, 81
  in our guts, 105
  total elimination of, 68–76, 81
  ubiquitous in humans, 82, 83
  and vitamin synthesis, 80–81
milk:
  adult ability to digest, 122–23,
    124, 126–27, 131–32, 135,
    136, 137–39
  breast, 80
  and cheese, 126
  on food pyramid, 132
  and population changes, 127
mimicry, 199
mites, 212, 213
mitochondria:
  microbial DNA in, 75
  origin of, 7
modernity (progress), 17, 21
mole rats, 213, 214
mollusks:

predators of, 171–73
  Vermeij's study of, 170–71
monkeys:
  appendix of, 95
  births of, 151–52
  color vision of, 166, 169, 174–
    75, 179
  fight-or-flight response in,
    152–53
  grooming of, 220
  predators of, 149–50, 174
  RNA retrovirus of, 167
  sleeping together, 150
  and snakes, 168–69, 173–76,
    193
  snoring, 150
mosquitoes, 214, 220
Murray, Damian, 222, 224
mutant genes, 121
mutualists, 30, 68, 81, 105
  animals and microbes, 85, 106
  and domestication, 124
  and evolution, 138
  obligate partners, 67

naked-bear paradox, 210
Naskrecki, Piotr, 177
National Bison Range, Montana,
    24–25
National Geographic, 9
natural selection, 26, 28, 79
nature:
  end of, 258
  human disconnection from, xi-xii
  persistence of life, 258
  restoring, 249
Neanderthals:
  extinction of, 11
  hairiness of, 203, 204
  tools of, 11
Nebuchadnezzar II, 240
Neel, James, 134–35
newborns, 80–81
New World, human arrival in, 12

New York City:
  green rooftops in, 240–42
  and hydroponics, 243
Niagara Escarpment, 253–54
Niger, snakebites in, 178
nits, 208
norepinephrine, 225
Nottingham, England, 19, 46, 56, 57
nudity, 213

obesity, 135, 228
Öhman, Arne, 194–95
Olmstead County, Minnesota, 19
Olmsted, Frederick, 244
oranges, osage, 29
overweight, 130, 133

Pagel, Mark, 207
panic, 162
Papua New Guinea, 213
parasites:
  absence of, 23, 32
  appendages of, 30
  and disease, 210
  ecto- vs. endo-, 208
  and hair/fur, 205–9, 211–12
  and immune system, 41–42, 56
  intestinal (worms), 21–22, 31–32
Parker, Bill, 98–104, 105
passenger pigeon, 61
Pasteur, Louis, 67, 72, 73, 75, 78, 84, 85
pathogens, 68, 81
peacekeepers, 42–43
peanuts, 118
penicillin, 62–63
Peru, Incan empire in, 234
pesticides, 61
phobias, 162
Picasso, Pablo, 97
pigeons, 61, 220, 248, 256
pig nematodes, 36–38

pigs, germ-free, 37
place cells, 195
plague, 213
plants:
  and agriculture, 117–18, 131
  evolution of, 40
  extinct, 131
  hydroponic, 243, 246–47
  interactions in, 102
  toxins in, 30
polio, 20
pollution, 258
pornography, 213
post-traumatic stress disorder, 162
potato blight, 128
predators:
  extinct, 27–29
  force exerted on prey by, 172–73
Pritchard, David, 56
probiotics, 250
progress:
  anticipation of, 17
  certain species favored by, 21
  sickening, 20
pronghorn (Antilocapra americana), 23–29
  populations of, 24
  predators of, 27–28, 62
  and rewilding the West, 34–35
  speed of, 23–24, 26–27, 28, 29, 32, 34
  uniqueness of, 23, 25
pronghorn principle, 30–31
proteins, 189
protists, 77
public gardens, 244

rage, 161, 162
Rantala, Markus, 207
rats, 248
  germ-free, 68–76
  grooming of, 220
  and snakes, 195
reason, 217